I0819106

The HIDDEN NATIONS *of* ANIMALS

The HIDDEN NATIONS *of* ANIMALS

A Grand Tour of Earth's Wild Civilizations

RYAN HULING

Maps & Illustrations by
OLIVER UBERTI

Avery
an imprint of Penguin Random House
New York

AVERY
an imprint of Penguin Random House LLC
1745 Broadway, New York, NY 10019
penguinrandomhouse.com

Library of Congress Cataloging-in-Publication Data has been applied for.
ISBN 9780593716847
Ebook ISBN 9780593716854

Printed in China
10 9 8 7 6 5 4 3 2 1

The authorized representative in the EU for product safety and compliance is Penguin Random House Ireland, Morrison Chambers, 32 Nassau Street, Dublin D02 YH68, Ireland, https://eu-contact.penguin.ie.

For Nellie—

May our adventures never cease.

Contents

Prologue

ONE LONG CENTURY AGO, a young Harvard graduate named Henry Sheahan was returning home from his time as a volunteer ambulance driver in World War I. Tormented by memories of tear-gas canisters and blown-up limbs, he sought refuge in writing, penning nonfiction stories and fairy tales under the name Henry Beston. While on assignment for a magazine called *The World's Work*, he was sent to Cape Cod to document the lives of Coast Guard officers on the open sea. During this time, far-eastern Massachusetts was not the idyllic getaway it is today, but a place that inspired abject terror among boat captains. Between the 1600s and the 1950s, more than three thousand shipwrecks occurred along the coast. To live on Cape Cod was to repeatedly witness humanity's unsuccessful attempts to triumph over the elements.

Something about this nowhere land appealed to Beston, who decided after completion of his article to stay there indefinitely. He leased some land along the beach near a town of a few hundred people and built a tiny cottage where he could focus on his writing and enjoy some solitude.

As Beston soon realized, solitude is not always as it appears. In fact, despite the small human population, Cape Cod was densely inhabited by a wide variety of seabirds, foxes, deer, and other individuals who slowly revealed themselves to him through his ten wood-framed windows.

Freed from the petty distractions of wartime life, Beston began to meticulously observe and document his nonhuman neighbors, whose daily routines appeared far more orderly and sensible than those of his fellow humans. In contrast to the instability and recklessness he saw in Europe, other animals appeared to behave in a logical manner; one that was thoughtful and reflective of changes in their environment. Theirs was a world of balance as ours was increasingly falling off its axis.

Writing in what would become his seminal work, *The Outermost House*, Beston presciently intuited from his seaside shack that the creatures we had traditionally seen as fleeting ephemera actually possessed far grander levels of organization and a quiet wisdom.

> We patronize them for their incompleteness, for their tragic fate of having taken form so far below ourselves. And therein we err, and greatly err. For the animal shall not be measured by man. In a world older and more complete than ours they move finished and complete, gifted with the extension of the senses we have lost or never attained, living by voices we shall never hear. They are not brethren, they are not underlings; *they are other nations, caught with ourselves in the net of life and time, fellow prisoners of the splendour and travail of the earth.*

When I first read this quote almost two decades ago, I was struck by the elegance of viewing other species not as accessories to our own lives, but as autonomous nations unto themselves. Enchanted by an inclusive vision of a world filled with adjacent, unseen animal civilizations.

As compelling as I found Beston's metaphor to be, I was acutely aware that the atlas of animal nations, to the extent that such a thing has ever existed, is drawn in invisible ink. The relationships, resources, conflict zones, knowledge hubs, borders, and landmarks that define nonhuman communities have always been recorded on what appear to us to be blank sheets of paper.

Writing these words from the second floor of my small 1920s cabin, built not on the beach but within a "wildland-urban interface" at the base of the San Gabriel Mountains, my mind flashes back to September 2020, when a massive wildfire incinerated more than 100,000 acres of nearby Angeles National Forest. I vividly recall gazing up at the colossal plumes of smoke rising from behind the ridgeline while live news coverage triumphantly declared that, despite widespread devastation, "no homes have been destroyed" and the blaze had caused "no injuries."

Those statements were plainly untrue, as countless forest residents, from rabbits and squirrels to coyotes and mountain lions, were killed or maimed in the megafire, and the areas they once called home were reduced to ash. But conventional thinking has long held that only human communities are inhabited places. Once we settle in a given area, it is, by definition, *populated.*

The town of Monowi, Nebraska, for instance, remains a fixture on state maps despite the fact that for the last twenty years it has had exactly one human resident.* By contrast, the millions of other

* Her name is Elsie Eiler. She serves as Monowi's mayor, treasurer, clerk, secretary, and librarian, and owns the Monowi Tavern, where she serves alcohol under a liquor license she issues to herself.

species we share this planet with have historically been viewed as living brief and transitory lives among pristine and *unpopulated* wilderness.

As our knowledge of animal societies and their relationships to the physical environment deepen, that distinction between populated and unpopulated places becomes increasingly untenable to draw—particularly when a fusion of scientific advancements is beginning to expose the invisible ink in animalkind's atlas to a new kind of heat.

IMAGINE, IF YOU WILL, a college student returning from summer break for the start of a new semester. Heading to her first class of the day, she pulls up the campus map on her phone and is intrigued to find two options: human or multispecies.

Instinctively, she clicks "human" and sees the familiar layout of the idyllic school grounds she's come to know. It confirms that she's headed in the right direction. She starts to put her phone away but hesitates. After pausing for a moment, she navigates back to her browser and clicks "multispecies."

Immediately, a vibrant new map unfolds, displaying not just the resident housing and sidewalks she recognizes, but also a galaxy of other color-coded markers. Seagull nests and flyways. Ponds teeming with frogs and newts. Hedgehog burrows and transit routes. A clickable kaleidoscope in the palm of her hand.

Locating the grassy quad she's standing on, the student zooms in on her favorite dining hall, the Stannary. Identified on the map by a blue fork-and-knife icon as a popular eatery, she notices that it's also tagged as a stopping point on the seagulls' flight path. Slowly peering up at the brick building ahead of her, she glances to the roof. Her eyes bulge when she spots a long row of birds, who appear to be having lunch. To nobody in particular, she blurts out, "No. Way."

This is the kind of scenario that animal geographer Dr. Sarah Crowley envisioned when she began work on a first-of-its-kind multispecies map of the University of Exeter's Penryn campus, in the far southwestern corner of the UK. Her map aims to achieve something as revolutionary as it is simple: reflect the housing, landmarks, and transit routes of all university residents, human and otherwise.

Touring the campus on an early-autumn morning with Crowley and the university's biodiversity officer, Abhishek Dixit, I was struck by the earthy overgrowth of the school grounds, which nature appeared determined to reclaim. As we walked through a carpet of burnt-orange leaves, glistening with a fresh drizzle of sea mist—*mizzle*, in local parlance—Dixit noted that some corners of the campus are classified as temperate rainforests; and they sound the part. The perpetual buzz of mining bees mingled with the shuffle of rabbits through the bushes, woodpeckers' rapid-fire drilling, and the hard-staccato *tchack, tchack, tchack* of ravenlike birds known as jackdaws, creating a subtle symphony.

At first glance, this quiet corner of the North Atlantic might seem an unlikely place to search for animal life. Nobody would confuse Penryn with the Amazon Basin or Serengeti as a wildlife hotspot, and at barely 100 acres, the campus is merely a satellite extension of a larger university system. But Penryn punches far above its weight in one notable regard: Nearly half of the university grounds are designated as green spaces, creating a landscape mosaic composed of thick forests, deep quarries, expansive grasslands, and dense marshes.

As we walked along a verdant path skirted by the compacted soil and stone of stately Cornish hedges, Dixit explained that identifying who is living on campus at any given moment has become something of a university pastime among the tight-knit populace of six thousand sustainability-oriented students. To date, more than 650 species have been found to call this place home—a figure we know because despite its wee size (or perhaps because of it), Penryn has become a living laboratory of animal monitoring.

For starters, it was the first institution in the world to boast a vertical-looking radar capable of identifying birds flying up to 4,000 feet overhead by analyzing the amplitude of their wingbeats. Every second, the radar can beam out 1,800 short-pulse radio waves in a massive cone shape. When the waves reflect back, the radar receives them and transmits data about the number, height, direction, and speed of aerial passersby.

At ground level, miniature GPS "backpacks" worn by hedgehogs provide ecologists with updates on the busiest thoroughfares used by Penryn's spiniest denizens. (Predictably, they favor the hedges.) Motion-activated camera traps scattered around campus offer field reports on red foxes, badgers, and other nonhuman visitors, while the geography department's high-resolution drone footage helps make sense of who or what is drawing animals to certain areas.

Eager to do their part, the university's ecology society, or EcoSoc, collaborates with other students to wage periodic "bio-blitzes," crowdsourcing comprehensive distribution maps of moths and other invertebrate residents. During a 2023 blitz, students discovered a completely new species of jumping spider, and then promptly found thirty more individuals from the same species across campus.

Around the time of my visit, the university also launched a new citizen-science app called Shell-E, which hundreds of students have used to log and identify the convoy paths of Penryn's thousand-plus snails as they painstakingly crawl across campus.

"The dream is to have most of the snails on campus individually marked so that when anybody stumbles across a snail and wants to follow their progress or see where they came from, they could go to the digital map," says Professor Dave Hodgson, the project overseer. Penryn may seem small to us, he notes, but to a snail, it's a seemingly endless world to explore. Hodgson's field research has revealed that snails have a homing instinct of up to 30 meters (about 100 feet). That may not seem like a lot until you consider that snails move at about one meter per hour, turning a journey across Penryn's walled garden into a multiday expedition.

In some respects, Penryn is the ideal setting for such a radical experiment to take place. The town is situated in Cornwall, one of Europe's six ancient Celtic nations—communities shaped by common descent, traditions, culture, or language. Surrounded on three sides by sheer cliffs, Cornwall's protruding peninsula faces the full brunt of harsh ocean gales, which sweep through the narrow cobbled streets of its many ports. This exposure has helped forge a distinct cultural identity that prioritizes the integration of human infrastructure with elements of the wilderness. The notion that nature begins only where humans end is anathema to the Cornish spirit.

The asserted "right to roam," which argues that land should be accessible to all rather than walled off into private parcels, looms large here, as it does across northern Europe. This mindset is perhaps best exemplified along the craggy Cornish coastline, where hundreds of miles of public trails have been thoughtfully preserved in a light-touch way. Eschewing fences and pavement in favor of natural footpaths lined with brambles and shrubs, these slender pathways provide epic sea views while leaving the ancestral homes of other animals intact.

To Crowley and Dixit, their multispecies map is a natural extension of this worldview by identifying places meaningful to Penryn's

A Multispecies View

The University of Exeter's Penryn campus is an interconnected mosaic of habitats and transit corridors for humans and nonhumans alike.

Rabbits and red foxes congregate in the open fields and shaded woods.

More than 300 moth species live among the wildflowers, grasslands, and thorny brambles.

Through the university's citizen-science app called Shell-E, more than a thousand snails have been labeled to pinpoint high-density areas and map their plodding movements.

In 2023, students found a completely new species—dubbed the "Tremough jumping spider," in honor of the historic estate the campus was built on—hiding in plain sight in roadside tree ferns.

Environment and
Sustainability Institute
The Stannary
Hedgehog
GPS tracks
A powerful rooftop
radar can identify birds
by wing-beat frequency.
Researchers tag resident
seagulls and jackdaws to
better understand their
needs and desires.
Jackdaw
colony
Old
badger
sett
Stream
OLD DRIVE
TRELIEVER ROAD
Star Pond
N
GPS tracking (red) shows that
Cornish hedges provide secure
nesting spots and hidden
highways for hedgehogs.
Rare Cornish yellow centipedes
make their homes in old logs.
Scale varies in this perspective.
Straight-line distance from
Star Pond to the Environment and
Sustainability Institute is 2,500 feet.

nonhuman residents and visitors. A blueprint that enshrines the right to roam for the entire animal kingdom.

SITTING IN HIS corner office at the Max Planck Institute of Animal Behavior in Möggingen, Germany, Dr. Martin Wikelski smiled approvingly when I told him what Crowley and Dixit are up to. If people are moved by the Penryn campus map, they aren't going to believe what he's got up his sleeve.

Wikelski leads a groundbreaking project called ICARUS, which stands for International Cooperation for Animal Research Using Space. ICARUS employs tiny tracking tags and powerful receivers installed on satellites 250 miles above Earth's surface to monitor and map animal movements, from antelope to falcons. Wikelski's program aims to provide, for the first time in human history, the ultimate bird's-eye view of what he describes as the "pulse of the living planet."

Initially tested through a proof-of-concept receiver on the International Space Station, ICARUS has since begun to scale up its operations by developing a series of smaller and more powerful microsatellites known as CubeSats. Each microsatellite, slightly larger than a Rubik's cube, can receive signals from at least 15 million unique animal ID tags once launched into orbit. When fully operational, Wikelski estimates, his ICARUS network will be able to monitor at least 40 percent of all bird species and half of all mammals on Earth—something akin to turning on fluorescent lights in a pitch-black room.

Even at this early stage, tracking technologies like those used by ICARUS are generating a fire hose of new data, most of which flows into a ballooning database called Movebank, also maintained by the Max Planck Institute. As of March 2025, Movebank has logged more

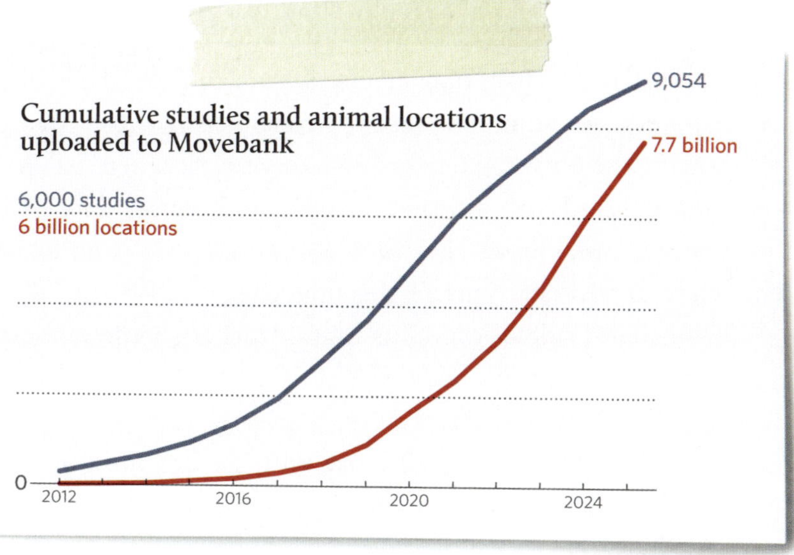

than 7.7 billion animal locations, spanning more than 1,500 species. At its current pace, Movebank adds nearly 12 million new data records every single day.

Wildlife monitoring is a field fraught with thorny ethical questions and trade-offs, but the ultimate goal is to better understand nonhuman societies with the lowest possible level of interference. As tracking technologies become cheaper and more prevalent, Wikelski is ever conscious of unintentionally becoming a burden to the animals he aims to learn more about. His team continuously redesigns lighter and more discreet tags that can deliver reams of information without inconveniencing their subjects. The private sector has also jumped into the animal-monitoring space. With the click of a button, researchers can now buy ultralight tags custom-built for squirrels, sloths, parrots, foxes, opossums, sharks, turkeys, flamingos, penguins, and nearly every other group imaginable.

For large terrestrial mammals, trackers often take the form of lightweight collars similar to a dog's. For smaller mammals, ear tags and harnesses are commonly used. In the case of marine animals like sea turtles and fish, trackers are affixed to their shells or skin using waterproof adhesives. When I first met Wikelski, his team was putting the finishing touches on a tiny tag designed to clip onto orangutans' beards in a way that the great apes won't notice. Within five years, he anticipates that they will develop transmitters light enough to unobtrusively attach to a honeybee.

The general rule of thumb is that a tracker should not exceed 5 percent of an animal's body weight, with an even lower threshold, 3 percent, for birds. There are no universal rules when it comes to insects, but one researcher I spoke with showed me a prototype of a minuscule solar-powered radio tag designed for monarch butterflies. Stuck to a piece of painter's tape to prevent it from floating away, the tag weighs 60 milligrams—comparable to a single grain of rice.

Even as animal tags get smaller, they are also growing much

more powerful. A decade ago, researchers would have considered themselves lucky to receive a single data point per day. A brief snapshot of how an animal has moved—or not—in the span of twenty-four hours. What happened in between those updates was anyone's guess. But now, those same researchers receive data pings once per second, depicted on high-resolution 3D maps that illustrate exactly how certain individuals move throughout their worlds.

To illuminate the difference, one Max Planck researcher told me that her earlier tracking work had revealed that the kinkajous she observes in the Panamanian rainforest return to the same fruit trees for food every day. That was an exciting discovery, but it became all the more intriguing when she started receiving geolocated, once-a-second updates showing that the animals not only visit the same food sources every day but also commute across the forest by walking along the exact same *branches*—a network of well-established aerial trails through the treetops. Suddenly, another puzzle piece in her mental portrait of the kinkajous' world fell into place.

AT THE SAME time that GPS trackers and tags are proliferating, other technologies are putting eyes and ears all over the landscape. Motion-sensor cameras and finely tuned acoustic devices have long served as valuable tools for capturing snapshots and audio clips of individual animals as they pass through a fixed area. Historically, biologists had to diligently review these recordings frame by frame, which limited how much information could reasonably be gleaned. But no longer. Now, artificial intelligence can handle much of the hard work for us, instantly identifying and logging everyone from hummingbirds to mountain lions by sound or appearance and pinging wildlife researchers' phones whenever their species of interest is spotted.

This shift toward automated observation is having a profound impact on how much we know about animal movements. When AI analysis showed that hedgehogs were constantly running into human-made fencing installed around Scottish rail lines, officers responded by cutting small holes in the fences, creating passages they dubbed "hedgehog highways." Anthony Dancer, a conservation specialist from the Zoological Society of London, has personally captured thousands of hours of audio and tens of thousands of data files from his research sites on British rail-network property, which have helped him identify all sorts of animals. "We couldn't have done it at that scale using human observers," he told *The Guardian*. "Only AI made it possible."

In northern Zambia, researchers used an array of GoPro cameras pointed at the night sky to capture forty-five hours of raw footage as African straw-colored fruit bats departed for their evening foraging jaunts. A trained AI tool then analyzed that footage to decipher just how many of the nocturnal mammals were soaring overhead. The grand total: 857,233 bats. It took the AI about fifty hours to complete that analysis, and manual spot checks showed that its detections were accurate 95 percent of the time. Not too shabby when you consider that counting individual animals on this same scale would have taken humans thirteen years.

Back in Germany, Max Planck postdoc Dr. Hemal Naik and his colleagues are pushing this line of research several steps further by using AI to not only identify certain kinds of animals but also make sense of what they're up to.

On the grounds of a repurposed eighteenth-century farm, Naik and his collaborators constructed a "scalable multimodal arena for real-time tracking behaviour of animals in large numbers," or SMART-BARN. At its center, his team built a human-scale black box theater encircled by dozens of 3D motion-capture cameras and microphones designed to localize sounds to specific areas of the room. But the most famous performers to ever grace this stage were not Hollywood starlets. They were pigeons.

Pigeons aren't particularly gifted thespians, but for Naik's purposes, that unexceptionality is a plus. All he needed them to do was act like birds.

One by one, and then in small, randomized groups, Naik brought in a total of eighteen pigeons, each outfitted with a small reflective sticker on their head and an ultralight backpack, akin to the green-screen suits worn by action stars. The positions of their beaks, tails, and other body parts were inferred from these sensors.

Then, the birds just hung out. For a total of about six hours over the course of more than thirty mini-sessions, they passed the time as one might expect: strutting around the room, sizing up their peers, interacting, taking short flights here and there, and pecking the ground looking for food. Pigeons being pigeons.

Once the stars were excused, the research team fed that raw footage into a machine-learning tool, which churned through the recordings and translated them into hyper-precise datasets. Every slight turn of a bird's head or stretch of a leg was meticulously recorded, resulting in millions of annotations, measured down to one three-hundredth of a second. The goal was to teach the AI exactly how pigeons move around their world and interact with their peers in various scenarios.

Then Naik put his tool to the test. He brought in the next group of birds, this time without any trackers or markers. Just regular ol' pigeons.

Remarkably, when he fed this new footage into his trained model, the AI not only recognized the birds as pigeons, but also instantly analyzed their body movements and behaviors, providing a detailed assessment of how the birds interacted with the space and each other. The kind of nuanced behavioral data that could have taken weeks to gather—with far less precision—through manual observation.

Encouraged by these early results, Naik and his colleagues decided to take their AI into the wild. By setting up a handful of cameras in a public park in Germany, researchers gathered footage of completely random groups of street pigeons who had never set foot in the SMART-BARN. The AI didn't even blink. It instantly recognized the pigeons, logged their body movements in exacting detail, and provided researchers with a comprehensive analysis of how these vagabond birds appeared to interact with each other and the landscape.

If such technologies are rolled out on a macro level, Naik says, humans will soon possess the ability to identify and study wild animals in a completely noninvasive way, and perhaps even assess the behaviors of entire communities.

Field researchers have already shown, for instance, that by flying camera-equipped drones over free-ranging animals in a particular area, such as a popular watering hole, algorithms can analyze the local landscape and ecosystem, identify the species seen, and sometimes also estimate the age and sex of the individuals below, effectively conducting an instant animal census. No tags, no traps—just eyes in the sky that know what to look for.

In some cases, AI has even demonstrated the capability to identify animals on an individual basis, so instead of merely labeling a striped mammal as a zebra, it can identify exactly which zebra it sees—nuances our eyes struggle with. Such tools can also determine whether animal groups are visiting an area to forage or if their movements are part of a well-established transit corridor.

This precision means that, in short order, we may be able to instantly assess not just who lives in a particular area and how they relate to their friends and neighbors, but also *why* certain locations are important to them. The landmarks—known, remembered, yearned for, and struggled over—that give animals what anthropologists would call a "sense of place."

Amid this technological renaissance, nonscientists have also been invited to participate in mapping animal worlds. Free apps like Merlin can instantly identify any bird from a captured picture or audio clip. AI identification has similarly supercharged citizen-science tools like iNaturalist, which has logged hundreds of millions of geolocated sightings, providing an unprecedented level of insight into where animals live and go. During the height of COVID-19, two roommates in an Australian suburb made international headlines by spending their lockdown days logging wildlife observations on iNaturalist. Within a year, they had successfully mapped the locations of more than a thousand animal species literally in their own backyard.

Leveraging crowdsourced sightings, some apps can even predict animal movements before they occur. A team of data scientists partnered with eBird, an app run by the Cornell Lab of Ornithology, to launch a tool called BirdFlow. Powered by observation records, and spot-checked by GPS and satellite data, this tool allows users not only to identify avian passersby they see out the window, but also to "connect the dots" by peeking into the future and learning where that bird is likely headed.

The simultaneous acceleration of these various technologies has brought humanity to a curious new place—one in which a mosaic of tags and tools is filling in the blank spots on our maps at a staggering pace, revealing clues about who lives where and why.

Wikelski of ICARUS told me that what initially motivated him to track animals from space was to monitor real-time disruptions in movement patterns—like an interspecies surveillance system. Peering wistfully across his office parking lot and into the thick forest beyond reminds him of how much in our world has, until very recently, remained a mystery. "There are thousands of deer in those woods. We have no idea what they're doing in there or how they're using that space."

By actively monitoring birds, fish, or mammals who are fleeing a given area, Wikelski says, we can look for causes, such as natural resource depletion, poaching, or disease. When his ICARUS satellites are fully scaled up, he believes humans may even be able to better predict phenomena that we don't fully understand, like earthquakes and volcanic eruptions, simply by noticing changes in how wildlife is behaving.

But the explosion of large-scale animal tracking is also gradually unveiling something more transformative: a reimagined version of our world in which humans are not the only species inhabiting any given physical space. A more scientifically correct and inclusive view of our surroundings that represents all known animal life, including humans, as overlapping and interconnected societies.

Around the world, wildlife experts and laypeople alike are watching with piqued anticipation as animals confound expectations, forging new pathways and building settlements in frontier areas far away from where we once believed they lived. We stand awestruck as animals congregate by the millions in what are, for all intents and purposes, bustling cities, complete with sophisticated transit networks. As new mapping techniques illustrate these hidden thoroughfares, terms like *highway*, once reserved solely for human infrastructure, are increasingly applied.

Every day, as new dots and zigzagging lines appear, the true contours of our world come into sharper focus.

THAT REIMAGINED WORLD is the one I will be traveling through in this book.

Empowered by our greater understanding of animal societies and in collaboration with leading ecologists and wildlife researchers, I decided to embark on a new kind of journey—a grand tour of Earth's wild civilizations.

I did so partly out of sheer curiosity. My professional guise as a sustainable food systems specialist has barely concealed a decades-long desire to explore the farthest reaches of the globe. Whether riding ramshackle railways across northern Myanmar or road-tripping between remote Patagonian villages, I have found no greater source of joy than an entrance into the unfamiliar. In my own home range of Southern California, I have been known to linger like a fly on the wall in anonymous diners, spontaneously ramble off for a day trip to a little-known ethnic enclave, or, better yet, spend a night under the Mojave's ocean of stars.

After so many years of ceaseless wandering, I had rather arrogantly come to believe that I possessed a broad understanding of the world we live in and the various communities that inhabit it. Jaunts to points unknown that might have once invigorated me began to feel homogenized, like visiting a new room in the same house. But as the nations of animals started to steadily materialize on our maps, redrawing borderlines and etching entirely new categories of landmarks into the finite world we knew, a healthy dose of humility rushed back into my head. A necessary reminder that there is still much to explore.

What would it look like, I wondered, to view places inhabited by other species not as escapes from society, but as opportunities to glimpse another? To treat nonhuman communities as sovereign spaces with their own histories, struggles, and unique relationships to the land, air, and sea?

My journey was also motivated by the knowledge that, historically, anonymity has done animals no favors. If we remain blissfully ignorant of the physical locations that matter to other species, it becomes much easier to inadvertently erase their communities and cultures from the earth. To prevent us from wiping animals off the map, they need to be on the map in the first place.

Each animal society looks distinctly different. Some are stationary, forming breathtaking metropolises in geographic places that can be precisely pinpointed. Others move along nomadic, Silk Road–style migration routes, but with clear affiliations between disparate individuals.

In an effort to demystify the animal settlements, cities, and nations I observed, I, too, will employ humanity's greatest visual tool: the humble map.

A well-designed map does not simply list the contents of an area, but answers the question "What *is* this place?" In his essay collection *Sunrise with Seamonsters*, novelist and travel writer Paul Theroux described maps as "a masterly form of compression, a way of miniaturizing a country or society." A map can do many things, he wrote, but "its chief use is in lessening our fear of foreign parts and helping us anticipate the problems of dislocation. Maps give the world coherence." Through my collaboration with renowned artist Oliver Uberti and use of cartographic themes in each chapter, I hope to bring that sense of clarity to the worlds of other animals, inspired by the profound sense of awe that accompanies an expedition into unknown lands.

In undertaking this odyssey, I knowingly ventured into murky waters. The staggering diversity within the animal kingdom means that other species see colors, hear sounds, and experience sensations we cannot. Even with the advantages of on-the-ground wildlife experts, advanced radio-tracking systems, and high-resolution satellite eyes in the sky, it was not at all clear whether the hidden nations of animals would fully reveal themselves to foreigners.

At a more fundamental level, the premise of my journey also posed a provocative question: If we, as a species, no longer divide the world between populated places and wilderness, are the wild places nowhere or everywhere?

1

Life on the Frontier

Settlements

THE MOST MEMORABLE journeys always begin with a bit of uncertainty. Uncertainty of how to reach one's destination. Uncertainty of what can be gained by doing so. In my experience, certainty on these points is the surest sign that a journey should not be undertaken.

As I abandoned my rental car where the roads end in northeastern Alberta and stepped into a tiny skiff to press on another 150 miles by river, I did not possess such reassurance.

Floating north along the thick veins of the Athabasca—"where there are plants one after another," in the Woods Cree language—the clattering of human society faded into the distance, only to be replaced by the percussive hum of other animal worlds.

I was joined on the water by residents from the tiny hamlet of Fort Chipewyan, Alberta's oldest human settlement. Three weeks earlier, the entire town—affectionately known as Chip—emptied out after a series of lightning strikes ignited dozens of simultaneous wildfires across the unusually dry boreal forest. Fast-approaching flames soon forced the remote outpost to issue the first mandatory evacuation since its founding in the 1700s.

Escaping Chip is no easy feat. In wintertime, an ice road offers a brief lifeline between the community and the rest of Alberta. More than half the town's fuel supplies are imported during this narrow window. By late spring, bush planes and boats are the only options left. Seeking safe haven, most evacuees fled to the small Indigenous community of Fort McKay, which by boat can be a full day's travel south, or pressed on a bit farther to the grimy oil-sands town of Fort McMurray.

Coincidentally, the very day that Chip residents started pouring south, I was in Fort McMurray, headed the other way. Sporting a waterproof backpack stuffed to the gills with survival tools, I was on my way to meet up with a Parks Canada ecologist to explore a remote corner of Alberta's northern interior. In anticipation of primitive conditions, I had a water filter and pump, heavy-duty hiking boots with metal lugs for trekking in muck, chest waders, a fly-resistant hat, headlamps, bear horns, a GPS tracker, and two weeks' worth of dehydrated food, among other essentials. The ecologist, whom I was planning to meet in Chip, was to provide the tent, cooking stove, first aid kit, and satellite phone.

Then, in a matter of hours, those plans completely unraveled. All onward flights to Chip were canceled indefinitely, and my northward progression came to a screeching halt amid Fort McMurray's strip malls and vape stores. Armed with exactly the wrong supplies, I looked up at the hazy-brown sky, which was filled with soot and toxic particulate matter—a signal of voracious charring on the horizon. It was the tail end of what Alberta would soon declare the hottest May on record. In previous years, around 1,000 acres would have burned by this point in the fire season. This year, it was 1.5 million and counting.

Fort Chipewyan's eight hundred Indigenous residents spent the next three weeks hunkered down in seedy hotels and public parks, where they muttered anxiously as daily updates dispatched by exhausted front-line crews warned of "extreme fire behaviour," "substantial growth," and "strong winds."

Only after nearly a month in exile did the good word finally emerge from the chief of the Athabasca Chipewyan First Nation: "We will be going home soon."

While wildfires continued to rage uncontrolled across the broader region, most residents did not want to spend another minute away from home, even if it meant heading back toward the flames. At my request, one resident, whom I'll name Frank,* was gracious enough to let me hitch a ride.

Chip was the final destination for Frank and most other people riding the river, but for me it was a gateway to something grander: a 1,100-mile string of settlements that, by most accounts, do not exist.

GAZING OUT THE left-hand side of our boat, I caught my first glimpse of the rugged eastern perimeter of Wood Buffalo National Park—a land saturated in superlatives.

Straddling Alberta and the Northwest Territories and covering an area bigger than Switzerland, this is Canada's largest national park and the most expansive protected wilderness zone anywhere in North America. It's home to the planet's largest free-roaming herd of wood bison, who are themselves the largest land mammals in the Americas. (The park's name stems from European explorers mistaking the bison for buffalo, a similar-looking mammal found in Asia and Africa.) Musk oxen, bears, cougars, wolverines, falcons, lynxes,

* Transporting outsiders was frowned upon amid the reentry chaos, so I have changed his name.

and many other iconic species of western lore have roamed the park's fertile lands for millennia. A visiting Frommer's writer once remarked, "Most of the animals here could easily live their entire lives without knowing humanity exists."

Maps are a fluid thing in a region like this, where shifting water flow and erratic weather mean navigable routes are there one day and gone the next. On no fewer than a dozen occasions, our boat was comfortably cruising through places where my phone's downloaded map said there should be land.

Beyond the imposing stockade of white spruce and balsam firs visible along the riverbanks, a good portion of Wood Buffalo is filled with dense peat bogs called muskeg—a seemingly endless quilt of stagnant pools, waterlogged vegetation, and organic material in various states of decomposition. This malleable landscape, created through the retreat of melting glaciers, is virtually impassable by humans, but provides ideal conditions for other species to build on.

As Frank and I made our final approach into Chip, our boat was spit out of a narrow tributary and tossed from side to side by violent gusts skidding like hockey pucks across Lake Athabasca. After more than ten hours meandering through relatively calm backwaters, we suddenly found ourselves assaulted by angry whitecaps, the silhouette of downtown faintly visible in the distance.

The water level was unusually shallow, as an acute drought had deprived locals of the springtime snowmelt that makes transit passable. Anticipating this challenge, Frank drew a squiggly red line on our downloaded map that cut a circuitous path across. If we deviated from this precise route, we risked scraping the lake bottom and becoming stranded. Staring at the screen, I shouted directions as Frank made incremental turns of the steering wheel.

After half an hour of exacting zigzags, our route finally passed through a string of small barrier islands—a welcome moment of tranquility that signified our unofficial entry into the heart of the region I had come to explore. A landscape expertly engineered not only by humans, but by beavers.

SETTLEMENTS ARE COMMONLY understood to be places of fixed inhabitation, often motivated by a strategic location, perceived absence of safety threats, or close proximity to natural resources. It's a fraught term frequently hurled at those seen as invaders, but at its most basic level, a settlement is simply a place where at least one person lives.

Yet despite beavers and other nonhuman animals having individual *personalities*, we have not historically viewed them as *persons* capable of physically occupying a place. When kindergartners are taught about the difference between a person (their teacher), a place (their classroom), and a thing (their desk), it's generally left unsaid which category animals fit into. As a result, settlements are broadly assumed to be beyond their capacity.

When humans decide whether or not to settle in a new place, we often take some basic criteria into account. Will we have reliable sources of food and water? What are the limits to how far we can travel? Perhaps a different group of humans (or other animals) lives nearby that we know to be aggressive. Are the risks of raising our family near that potential threat outweighed by the advantages this area provides to us? And if we do decide to stay, who's going to do what in order to make sure that our community thrives?

These are manifestations of what philosophers refer to as *worlding*—the process of creating one's own world. It involves a series of steps, such as identifying landmarks, drawing spatial boundaries, assessing the ecological context and vital contributions others are making to it, and establishing our interactions with the fabrics of the planet, which collectively form *the* world as a group of individuals understands it. These life-sustaining practices continually shape and reshape the narratives and mental maps we carry with us everywhere we go—and a growing number of experts now believe humans are not alone in enacting this process.

"Surely there has to be a way for all animals to consistently sustain and reproduce themselves, behave predictably, and all of the things that we really want to do when we make worlds," says Dr. Greg Anderson, a historian of ancient human civilizations at The Ohio State University. "World-building may be easier to see when humans or other animals gather in large numbers or organize themselves in a highly coordinated way like an ant colony. But even in cases like a wolf pack, they have their own model of the world. They live in small, nuclear-type families, so their world might look pretty limited to us, but certainly it would involve paths here and there, and rewards that will be found in certain reliable locations. They have to have an understanding of what keeps a wolf alive, how and where to reproduce, and what sustenance can be gained by following a particular practice. Are we really going to claim that wolves don't have practices? Of course they do."

Dr. Meg Crofoot, an evolutionary anthropologist at the Max Planck Institute, agrees. "All animals have a clear investment in the landscape, even if they're not changing it as dramatically as a beaver or an elephant might. Locations that have particular importance, or that relay information exclusive to that place. That information is inherently valuable because it provides security and helps animals to make sense of the world as they see it."

To skeptical audiences who are quick to brush off such comparisons as anthropomorphism run amok, Anderson reminds people that they have it exactly backward. "Humans are animals too, and we engage in the same classic worlding behaviors that help all species understand their environments, adapt, and survive. Humanity is not the custodian of creation. Indeed, we're often the followers, chasing other animals and learning how to navigate their worlds, which we then think of as our own."

THE FAR-FLUNG EDGE of Alberta I was pulling into represents the northwestern corner of what Canadian ecologist Jean Thie has dubbed North America's "beaver belt."

For more than fifty years, Thie has been at the forefront of large-scale environmental management and monitoring in Canada, beginning with his aerial assessments by helicopter and fixed-wing aircraft in the 1960s. In his work for the national government and various nonprofit organizations like the International Union for Conservation of Nature, Thie has made a career of scouring boreal and Arctic regions for notable patterns and trends in environmental change, particularly permafrost melt.

Among the most notable events in our planet's history, Thie says, was the melting of glaciers at the end of the last ice age, roughly ten thousand years ago. Across the continent, ice sheets, some more than a mile thick, gradually warmed and thawed, turning the ice into meltwater and forming extensive glacial lakes. Over time, the lakes receded, radically reshaping the earth and creating sand-and-gravel shoreline ridges hundreds of miles long. As the lakes shrank further, extensive wetlands emerged inside the former lake beds.

As Thie studied these faint topographical lines more closely, a clear trend started to emerge: The slightly sloping wetlands in his study area, just below the glacial shorelines, were all populated by beaver dams. Huge ones.

He did not initially consider this discovery to be of significant scientific importance, and beavers were not his focus at the time, but it did pique his curiosity. Beavers are found all across North America, but the noticeably higher concentration in his research areas struck him as a trend worthy of further exploration. Having served as director of the world's first geographic information system (GIS), which debuted in Canada in the 1960s, he also knew that, through a combination of raw satellite imagery and historical data from his own field observations, it was at least theoretically possible to discern what factors were driving the beaver construction boom he glimpsed from the air. Not easy, but possible.

Then, in June 2005, a digital tool emerged that promised to simplify and democratize Thie's plodding work. "When Google Earth launched, I just assumed a lot of people would start looking at long beaver dams for the hell of it. So I figured I should provide some instructions on how to do it." From the basement of his log cabin outside Ottawa, Thie toggled back and forth between four computer monitors and typed out a series of detailed guidelines for using this new tool to look for signatures of animal life from the heavens, in the hopes that many hands would make for lighter work.

Thie was overly optimistic. After a brief halcyon period when like-minded community users joined him in scouring the remote corners of Google Earth's archives, Google Maps debuted on mobile devices, Street View–style navigation became all the rage, and for most people, satellite images took a back seat.

Undeterred, he continued to throw himself into the work, using Google Earth to "revisit" distant fieldwork areas he'd toured in person decades earlier to see how they'd changed. What a shame that others moved on, he thought, because perhaps more than any other nonhuman habitat, beaver dams are particularly well suited to being viewed from the stars.

- - - - - - - - - - - - - - - - - - -

WHEN BEAVERS ARE scouting a possible location for their dam, one of the first points they consider is where they hear flowing water. Gentle creeks and rivers are more appealing than, say, white-water rapids, which would swiftly overpower them. Thie also noticed that given the choice, beavers are partial to places just downstream from plateaus and foothills, because in the springtime, snowmelt from higher elevations brings with it huge bounties of runoff sediment, mud, and sticks. For a beaver, that's like having their groceries and home-maintenance supplies delivered to their doorstep.

Once a suitable location has been selected, construction can begin in earnest. Aided by up to a half dozen family members, the beavers will harvest branches and logs from the surrounding forest under cover of nightfall, carrying and shoving them one by one into the floor of mud and rock. This structure is then interwoven by a dense mesh of twigs and vegetation, which plugs holes and creates a solid barricade capable of retaining water and flooding the landscape. This hydro-engineered beaver pond forms the foundation of their world.

As the water rises, the beavers will select a spot, usually in the center of the pond, to build a lodge. The water around the lodge has to be deep enough to form a moat, providing defensible space from predators who might wish them or their family harm. Made of branches and mud, these domed structures will ultimately serve as their primary housing, complete with secret tunnels, underwater doorways, and several interior chambers for living and sleeping.

Beavers don't hibernate, preferring instead to spend autumn as many human settlers do: cutting and storing wood in anticipation of leaner times. Food is also stockpiled underwater near their lodge, so that even when the pond freezes over in the depths of winter, they can swim out and grab a quick bite to eat. Sometimes a second den will be created on the riverbank, just to have options.

Astrophysicist Neil deGrasse Tyson once remarked how familiar this process should sound to our ears. "'Oh, there's a tree. I'm going to use that tree to dam this river and I'm gonna make an underground den.' All right? Are we any different from that? We use trees. Well, first we used grass to make huts; that was available. Then we used trees. That's pretty convenient. Then we found metal. 'Oh my gosh, let's use that!' OK, and then we learned how to make alloys. 'Let's do that.' And then we learned chemistry. 'Let's do that.' So yes, it takes thresholds of intelligence to exploit your environment even more, but the simple act of exploiting an environment is not unique to being human."

Once constructed, beaver dams are remarkably sturdy. When researchers took a fresh look at an 1868 map drawn by anthropologist and politician Lewis Henry Morgan of what he described as "a beaver district" in Michigan's Upper Peninsula composed of sixty-four dams and ponds, surveyors found that 75 percent of the dams were still standing—or perhaps had been blown out and rebuilt in the same spot—150 years later. The dams, Morgan wrote, "have existed in the same places for hundreds and thousands of years," and "have been maintained by a system of continuous repairs." In Canada's Yukon territory, recovered pieces of fossilized beaver dams have been dated to 125,000 years old.

Equally important to the beavers' survival is the creation of an intricate maze of navigable canals, which can extend hundreds of feet into the surrounding forest. Wood is much easier to drag in water than on land, so these channels create glide paths to replenish their supplies. Such infrastructure is largely invisible on modern-day maps because "canals" have historically been equated only with man-made creations. From a cartographer's perspective, animal-made canals may as well have been birthed by nature itself, like a creek, when in reality both versions were painstakingly dug with transit in mind.

If the beavers' efforts are successful, the result is a secure year-round homestead for their family. A living testament to the fact that for many animals, home is a place created, not found.

WHILE BEAVERS ARE perhaps the most visible example of animals engineering the physical world to their liking, they are not alone.

Muskrats commonly build lodges similar to those of beavers, but with foot-thick walls made of grasses and wetland plants instead of sticks. Indeed, muskrats will often tag along when beavers reengineer a landscape, joining other temporary or permanent pond residents like frogs and insects to create a sort of multispecies neighborhood.

In Central Africa, goliath frogs—the world's largest, weighing up to seven pounds—have similarly been found to push gravel and stones around to dam waterways and create peaceful ponds where their eggs and tadpoles will be safe from predators and rising water levels. Cathedral termites in Australia build mounds more than 15 feet tall, which, relative to their individual size, makes their buildings significantly larger than humanity's tallest skyscraper is to us. Gopher tortoises dig burrows that can stretch more than 50 feet long, at depths of two dozen feet below the earth's surface. And in the Kalahari Desert, sociable weaver birds construct enormous haystack-looking communal nests made of stiff grasses and a twig-and-turf roof, which can house up to four hundred residents and contain up to one hundred individual rooms. White-browed sparrow weavers, by contrast, work together to construct non-communal nests, where they roost individually and breed. But research published in the journal *Science* shows that their nests are all built according to the distinctive style of their cultural group, each of which has its own architectural traditions.

Such ecosystem engineering has been underway for the entirety of our existence as a species, but only in recent years have the results started to become visible to us on a macro scale. In the years following Google Earth's initial launch, the quality of satellite imagery accelerated rapidly. Whereas digital sleuths like Jean Thie could once zoom in only on pixels the size of a baseball diamond, recent upgrades now enable us to spot individual objects the size of a sheet of loose-leaf paper. At the time I am writing this, a US company has received preliminary approval to zoom in even further, showing pixels as small as four inches across—roughly the size of a deck of playing cards. For people looking for signs of animal life, that's a seismic shift. An entire universe of well-defined trails and habitats, previously lost in a blur of pixels, could soon be visible, clear as day.

Even before these hyper-precise eyes launched into the sky, Thie was plenty busy. With his beaver formula in hand, he started scanning across vast areas of topographical recession. He looked in remote corners of Finland, Russia, Belarus, and Tierra del Fuego. But his biggest discoveries were found in the areas he thought he knew best: the basins of Canada's glacial shores.

In October 2007, Thie stumbled onto what was soon declared the longest beaver dam on Earth, hidden among the spruce and firs of Wood Buffalo National Park. The dam is 2,700 feet across—the length of seven football fields. From the ground, it appears in photos to simply be a generic large pond; but when viewed with high-resolution cameras from above, the canals and family lodges are visible in all their glory.

Upon identifying the pond as an intentional beaver creation, Thie started looking back at archival satellite imagery to gauge roughly when the dam was built. He saw no record of it in aerial shots from the mid-1980s, so he skipped ahead to more recent data. 2006? It was there. 2005? Already there. 2000? There. Year after year, as far back as at least 1990, it was there. Nearly two decades before

he found it, the ambitious beavers' family settlement had been there, hiding in plain sight.

Once the media caught wind of Thie's discovery, his phone began ringing off the hook. He was interviewed live on news stations around the globe to talk about what he'd found. With characteristic understatement, he often described the moment as a kind of nonevent. "It's not the prettiest or most impressive dam to look at—it's just very long."

He was also convinced that there was much more out there. Not far from the world's longest dam, he had already found many others of exceptional lengths spanning more than 1,600 feet. Along the Alberta portion of the Peace River, he identified what he believes is the largest landscape entirely shaped by beavers—spanning more than 600 square miles. Beavers, he said, act as gatekeepers to this region, controlling every river and stream.

This vast province, including the forest surrounding the world's longest dam, was now engulfed in flames.

WITH MOST OF Fort Chipewyan's evacuated residents still en route home, Frank and I pulled our boat into "Big Dock" harbor, which was vacant except for a smattering of bulldozers and rusted husks of decommissioned vessels. As I walked toward one of the settlement's few guesthouses, I passed through a small cemetery next to a run-down Anglican church. Faded gravestones dated back to the early 1800s, emblazoned with familiar surnames: Flett, Wylie, Loutit. The same names that appear on the street signs.

Walking around, I noticed that most restaurants and businesses were still boarded up except for the Chief's Corner convenience shop and a grocery store advertising Beaver Buzz energy drinks ("dam good"). Stop signs were written in three languages: English, Cree, and Dëne.

The town now known as Chip falls within the ancestral territory of the Athabasca Chipewyan people, whom archaeological analysis shows were roaming across the region for more than ten thousand years before European traders ever set foot in this part of the world. When Europeans did eventually show up, they got there by navigating the same labyrinth of gradually sloping rivers and wetlands that make this area so ideal for beaver construction.

Prior to the Europeans' arrival, there were anywhere from 100 to 400 million beavers in North America—a veritable galaxy of animal settlements. Even years later, in his book *The World of the Beaver*, wildlife photographer Leonard Lee Rue III recounted seeing dams up to 18 feet tall on travels through Wyoming. In the early 1800s, Lewis and Clark's westward expedition passed a beaver dam upstream from what is now Three Forks, Montana, which at one point stood at more than 15 feet high and 24 feet thick. Farther east in New Hampshire, a single beaver dam reportedly stretched 4,000 feet long and contained forty individual family lodges behind it.

This abundance was not lost on the European traders, who pushed into Alberta's interior with a singular focus top of mind: hats.

At the time, the most highly sought-after fashion accessory in Europe was a felt top hat—an item that instantly projected wealth and status. Animal fur was the primary material used in its production, and within the fur world, beaver was king.

Owing to the beavers' preference for cold aquatic areas, their fur was said to be naturally warm and water-resistant. But even by comparison to beaver pelts from elsewhere on the continent, those from this corner of North America were reputed to be the finest in all the land. At the Montreal depot, they could identify Athabasca Delta fur just by looking at it, according to Maureen Clarke, director of the Fort Chipewyan Bicentennial Museum, which is filled with gunpowder flasks, snowshoes, travel trunks, colonial-era sewing machines, taxidermied fish, and other historical artifacts. "It didn't need to be tagged for them to know where it came from."

The Chipewyan people, like countless other Indigenous groups, have a complicated relationship with the beaver. Many First Nations tribes view nonhuman worlds as independent communities, sometimes referring to beavers and other creatures as "the animal people." As writer Leila Philip noted in her book *Beaverland*, citing research by historian Calvin Martin, every animal "functioned in a society that was parallel to that of mankind. And each species had its leaders, called 'keepers of the game,' or 'animal bosses.' . . . If a hunter angered an animal boss by disrespecting his people"—through overhunting, for example—"he ran the

risk of having his next hunt ruined, or of falling ill." When diseases spread among Indigenous communities after first contact with the European visitors, some believed this was a sign of collective punishment. "The response was rage; if the beaver would so abandon them, they would abandon their traditional protocols and hunt them without limit."

In the ensuing decades, the Chipewyan people and other Indigenous communities proved to be vital partners in exploiting the continent's abundance of wildlife, supplying Europeans with food and manufactured goods like snowshoes and sleds in exchange for guns and hunting knives. Indigenous middlemen also facilitated fur trapping and trade in far-flung regions that Europeans could not reach on their own. Indeed, beavers soon became one of the most commonly illustrated animals depicted on world maps because their very visage was synonymous with economic opportunity among European colonizers.

Among the earliest beneficiaries of Indigenous hospitality was Peter Pond, a founding member of the North West Company. Historians who have studied his letters say that Pond reportedly "procured twice as many furs as his canoes would carry" in a single winter season in 1788.* Chip was soon declared the new "Emporium of the North" and became the richest trading post in all of North America.

For two hundred years, beaver homesteads were systematically pillaged across the continent, pushing the animal people to the brink of extinction.

FROM MY CHIPEWYAN guesthouse, I closely monitored the latest updates on the wildfire situation by chasing little red dots around the NASA satellite map. Within the park limits, fifteen separate fires were actively rampaging through the forest, cloaking Wood Buffalo in smoke.

Under siege on all sides, beleaguered firefighting crews had triaged their resources to protect the human infrastructure. When the blazes initially ignited on Chip's outskirts and residents were ordered to flee, hotshot crews carved firebreaks hundreds of feet wide into the earth and dropped more than 35,000 gallons of flame retardant. Across Alberta, reinforcements arrived from as far away as South Africa and New Zealand to join the province's containment efforts.

But a few dozen miles west of Chip as the crow flies, no such help was on the way to the world's longest beaver dam. On the Parks Canada fire tracker, many blazes were color-coded in either orange or blue, signifying that they were "being managed" or "being held,"

* Pond himself had a devious and violent reputation. After presenting a series of maps of North America to Congress, he was accused of deliberately altering his drawings so as to confuse rival traders. He was later implicated in the murders of two fellow fur traders. The scandal so tainted his reputation that he sold off his company shares, moved to Connecticut, and died a pauper.

respectively. The two fast-moving fires actively burning within nine miles of the dam—in an area where wildfires can scorch up to eighteen miles in any direction in a single day—were marked in yellow, for "observed."

Even as Chip residents made their triumphant return home, animals across the forest were initiating their own evacuation plans. In emergency situations, elk, moose, and deer will often try to outrun blazes to reach water, University of Alberta professor Erin Bayne told me, while smaller mammals like deer mice will ride out the danger in their underground bunkers. One former Wood Buffalo firefighter reported that when an inferno is headed their way, many bear cubs will instinctively run up a tree in a desperate attempt to avoid harm. Their mother knows better but will often refuse to abandon her kids, causing the whole family to perish.

Those who do make it out of a fire's path alive often find themselves indefinitely displaced and disoriented, with none of the familiar geographical landmarks they have come to know. When a forest burns to the ground, vegetation can take up to a dozen years to return to the levels that large mammals need. Their world, as they knew it, is gone.

TO WITNESS FIRSTHAND the devastation sustained across the greater Wood Buffalo area, I drove out of town in a pickup truck loaned to me by a local security guard. Just past the airport, still closed to passengers, I followed a dirt road a few miles deep into the woods before reaching a demarcated line where the forest abruptly changed from a dense warren of noble greens to a postapocalyptic hellscape of puny black twigs. The road ahead was blocked by a tangle of ashen extrees. Beyond them, a haunting breeze was the only sound.

Exposed to the elements in ways that insulated metropolises generally are not, remote settlements, human or otherwise, carry an inherent air of precariousness. Residents retain an acute sense of smallness that comes from the knowledge that with one severe drought, flood, or fire, tiny dots like theirs can be smudged out of existence.

To overcome these risks, the Alberta fire crews' handiwork was visible everywhere. Huge tracts of trees had been bulldozed to reduce potential fuels for the fire to burn. Sprinklers had been affixed to every house along the human settlement's perimeter. Using massive excavators, fire crews had carved deep into the earth along the shores of Lake Athabasca to dig canals that could serve as natural firebreaks between the forest and the outskirts of town—a page seemingly ripped directly from the beavers' playbook.

The all-of-the-above assault worked. Between the newly dredged canals and airdrops of retardant, only ten rural cabins were lost in the blaze, all of which were deep in the forest on tribal land. Their owners pledged to rebuild.

Meanwhile, inside the park, the beavers' fate was far less certain. For days, I helplessly refreshed online fire maps as red blobs flared, faded, and roared back to life, inching closer to the longest dam like an encroaching army circling its enemy. It was only a matter of time, I surmised, before this decades-old structure was erased from the map, to the extent it was ever on one. It felt deeply unjust, given that beavers did little to exacerbate the climate conditions fueling the megafires across the continent. Indeed, their widespread presence once provided a reliable backstop against such catastrophes.

Then, just as I braced myself for final impact, the fires approaching the dam suddenly halted their forward progression. The red dots, clanging like tiny alarm bells, started burning in place, static on the page. I restarted my browser, thinking that perhaps it was mistakenly pulling up old data. Or had the clouds of smoke created visibility issues for the fire mappers? That had been known to happen.

No, I soon realized, zooming in on the landscape. Barely six miles

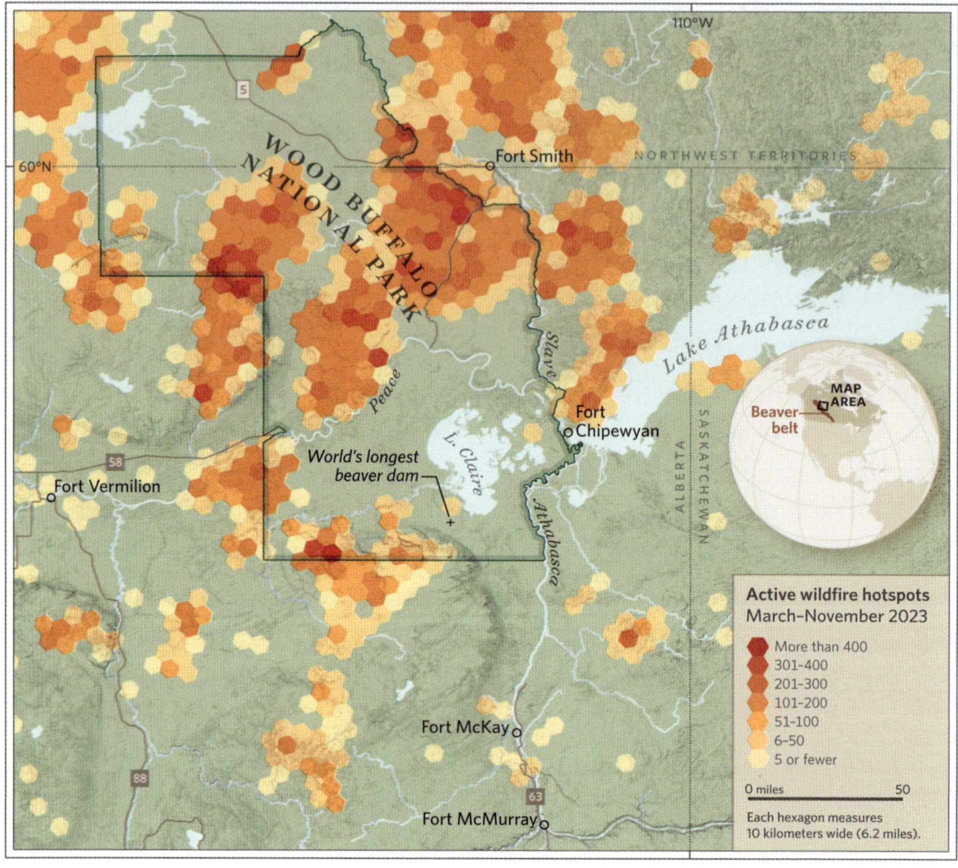

from the longest dam—a stone's throw away by wildfire standards—the blazes had run into a barrier: the wetlands carefully engineered by beavers. Frustrated by a lack of fuel and stymied by deep moisture on the ground, the flames retreated from their relentless march and pivoted toward drier terrain. At least for now, these beavers and their woodland settlements would endure.

While savoring that small victory on their behalf, pain pangs reminded me that Alberta's fire season was just getting underway. Every indicator suggested that the situation would worsen in the months ahead as the fires spread like cancer through the park. As it was, Canadian wildfire smoke was turning skies orange as far away as New York City. Burned-out Parks Canada staff soon decided that even though the summer backcountry season had just begun, Wood Buffalo would remain closed to all visitors until the snowfall eventually put the fires to rest that winter. "Maybe try again next year," I was told as bush plane operators announced the return of regular flights between Chip and McMurray.

Depressed and disgruntled, I sought guidance on how to move forward from the only person who might know: Jean Thie.

When I told him of the trials and tribulations I'd encountered attempting to explore the western portion of the beaver belt, Thie offered reassurance. While the world's longest dam got the eyeballs of the Guinness World Records crowd, in his opinion a far more impressive demonstration of beaver prowess could be found outside Wood Buffalo, hundreds of miles diagonally to the southeast, on the belt's eastern edge. There, he said, was a place far more awe-inspiring than any individual dam. An area that is truly "one in a million." If I wanted to close the loop and see beaver engineering in all its glory, I needed to get to Saskatchewan.

AS I WAS heading east on Highway 55, just past the midsize town of Nipawin, all commercial pop stations suddenly faded to static. They were replaced by a single option: "Saskatchewan's Indigenous Radio Network."

Fifty miles farther down the road, traffic came to an abrupt halt as a mother black bear ambled across the road with her two cubs in tow. A few minutes later, I took a sharp left off the main road and pulled into the thousand-person town of Pakwaw Lake, situated in the ancestral territory of the Shoal Lake Cree Nation.

The town itself looked like it had seen better days. Windows boarded up, paint peeling. Roofs in various states of collapse, and lawns overtaken by weeds. A run-down health center was clearly in desperate need of repairs. Pakwaw Lake's only store and gas station had both closed years ago.

Walking into the Nation's headquarters, I met with Chief Marcel Head, who bemoaned his town's precipitous decline. As in Chip, the beaver fur trade had once been a financial boon to this part of Saskatchewan, he said. For hundreds of years, most of the community made their living by trapping beavers and selling their pelts. But following pressure from animal activists, global fur demand had evaporated in recent decades. Some older residents pivoted to farming wild rice and other crops, but this agrarian lifestyle held little appeal to younger generations, who have increasingly left home for big-city life.

While he was recounting this history, a ragtag group of gruff middle-aged men arrived, dressed in camouflage and jeans. Carl, the team leader, was the chief's brother-in-law. Alongside him were his friends Roene, Eric, and Jerry. As with others in town, these gentlemen could no longer make a living from the fur trade, the chief explained, but they still knew the surrounding woods like the back of their hands.

In anticipation of my arrival, the chief had arranged for Carl to show me around for a couple of days. Not having much else going on, his friends decided to tag along. Looking at the rental car I'd been driving and sleeping in, Eric said, "Kia. In our native tongue, that means 'it's yours.'"

Some of the land immediately surrounding Pakwaw Lake is now owned by the Canadian government, but as part of a treaty, members of the Shoal Lake Nation have access to roam and harvest from it as they see fit. There are few concrete roads in this area, Carl said, though an ancient network of logging and hunting trails is still passable if one knows where to look.

We hopped on the back of a small fleet of 4x4 ATVs and drove down Main Street to the edge of town. Instead of turning to the right, where the road curves back toward the entrance gate, we went left onto an unpaved trail.

After riding along a muddy path that ran parallel to the town's drainage canal, we abruptly stopped at a spot where the stream suddenly forked to the left instead of continuing straight. Grasses had also been trampled, he noted, creating a narrow pathway through the reeds. "This was them," Carl said, pointing to a muddy slope that led from the stream into the adjacent woods. Carl and I climbed off our ATV and walked in our swishy chest waders along the pathway for about five minutes before arriving at an enormous felled tree, 30 feet long, lying on its side. The gnawed stump was still bright white inside, with shards of fresh wood chips scattered on the forest floor and green leaves on the branches, indicating this was recent work.

I asked Carl how long it would have taken a beaver to bring down something so colossal. "About half a day," he said, glancing stoically into the distance to survey the scene. They can't carry a tree this large, he noted. It's way too heavy. What they're really after are the twigs and branches higher up, which would normally be out of reach but have now been brought down to earth.

Circling back toward the main road, we swapped our ATVs for a pickup truck, drove a bit farther afield, and then pulled over at what appeared to be a stretch of uninterrupted forest. Stepping out of the truck, Carl pointed to a narrow footpath that led deep into the woods. The other guys stayed behind, but I followed as he pushed his way through a seemingly impenetrable wall of brush, occasionally glancing down at his Google Earth app as thorny branches latched onto his jacket, trying to hold him back. "Watch your step," Carl cautioned, pointing to a deep channel cut several feet into the earth. "You don't want to fall in there."

After twenty more minutes of tenacious trekking, we had managed to travel only a short distance, but when we looked horizontally,

the wide-open sky was strikingly visible between the trees. Making one final push through the brambles, I saw that the land suddenly opened into a vast, reed-lined lake a thousand feet across. And there, amid the reflective glow of the pine-tree backdrop, sat a large domed lodge.

"Come," Carl said. "There's more." He ventured down the trail to where the pond ends and pointed through to the other side of a brief barrier of trees and mud, where there stood a completely separate beaver pond—possibly larger than the first. This one has existed for at least thirty years, Carl estimated, like others nearby, as part of a patterned landscape that extends for miles in every direction. "We have hundreds of these," he said.

It's this boggy area that possesses what Jean Thie believes is the

densest concentration of beaver dams on Earth—the beaver capital of the world.

We returned to our truck and drove farther up the road before taking a sharp turn onto an unpaved logging trail cut deep into the forest. Revving our way uphill in all-wheel drive, we soon arrived at an end point beyond which we had to proceed on foot. My GPS tracker showed that we were 1,327 feet above sea level, standing more than halfway up the slopes of the Pasquia Hills.

These ridges, Thie says, are what make this remarkable area possible. As in the muskeg of Wood Buffalo, the relatively flat, slightly sloping topography of the terrain surrounding the hills has created ideal wetland conditions for beavers to settle in. During springtime, water flow is consistent but mild enough not to sweep them away.

Not coincidentally, Thie says, this hotspot of ecosystem engineering is situated on the buried shorelines of former glacial lake Agassiz—once known as "the sixth Great Lake" and larger than all the modern Great Lakes combined. He started doing helicopter fieldwork over the Agassiz permafrost in the 1960s and has seen enormous changes since then. While beavers were no doubt a large presence in and around Pakwaw Lake prior to the rise of the fur trade, aerial imagery from the 1940s and '50s shows a far reduced density. But roughly four decades ago, the beaver population started to reclaim its ancestral territory to such a degree that Thie now believes there are more beavers here than ever before—far outnumbering the human populations nearby. A comeback and then some.

So substantial has been the beavers' influence that in Thie's estimation, the area around Pakwaw Lake now no longer fits the conventional definition of a wetland. The animals' large-scale ecosystem engineering, he says, has created an entirely new kind of terrain that does not have any known parallels: a beaver landscape.

IN THE NEARLY twenty years since Thie first stumbled onto Pakwaw Lake, new tools have emerged that appear to validate his calculations.

Back when I was still holed up in my Chipewyan guesthouse watching wildfires ricochet around Wood Buffalo, a fascinating article titled "Mapping Beaver Dams with Machine Learning" hit my inbox. Published only a week earlier, it outlined how University of Minnesota geographer Dr. Emily Fairfax had worked with a team from Google to introduce a new machine-learning tool called the Earth Engine Automated Geospatial Element Recognition model, or EEAGER.

Armed with thirteen thousand satellite images featuring confirmed beaver dams, along with some fifty-six thousand dam-less locations, Fairfax and her team worked to train EEAGER to accurately identify what is and is not a beaver dam from above.

The training process had its fair share of early hiccups. The cul-de-sacs of American suburbs presented an unexpected challenge, Fairfax told me, because to the AI's eye, the asphalt circles surrounded by sidewalks looked a whole lot like beaver ponds. After refining the AI with a series of carrots and sticks, EEAGER was soon able to better distinguish between human and nonhuman constructions, achieving a remarkable 98.5 percent accuracy rate.

Knowing of Jean Thie's predictions, I asked Fairfax if she would consider enlisting her team's help to look more closely at Pakwaw Lake and do their best to illuminate the scale of beaver building taking place there. She agreed to give it a shot.

The results were staggering in scope. In this relatively small corner of Saskatchewan, her expert team personally identified more than 2,700 individual dams, which she speculated were home to at least two thousand families. Each family, Fairfax estimated, probably had between two and ten beavers in it, meaning that even the most conservative estimate would put the total beaver population at four times that of the human settlement next door.

FOR THE NEXT two days, my guides and I developed a familiar rhythm. We would pull over, march our way through a thick curtain of trees and brush, and stand on the muddy banks of vast ponds admiring the beavers' handiwork, which left virtually no part of this region untouched.

Like Fort Chipewyan residents, beavers are constantly monitoring the local water levels. They know their ponds so intimately, Carl said, that even the slightest change in depth can be immediately

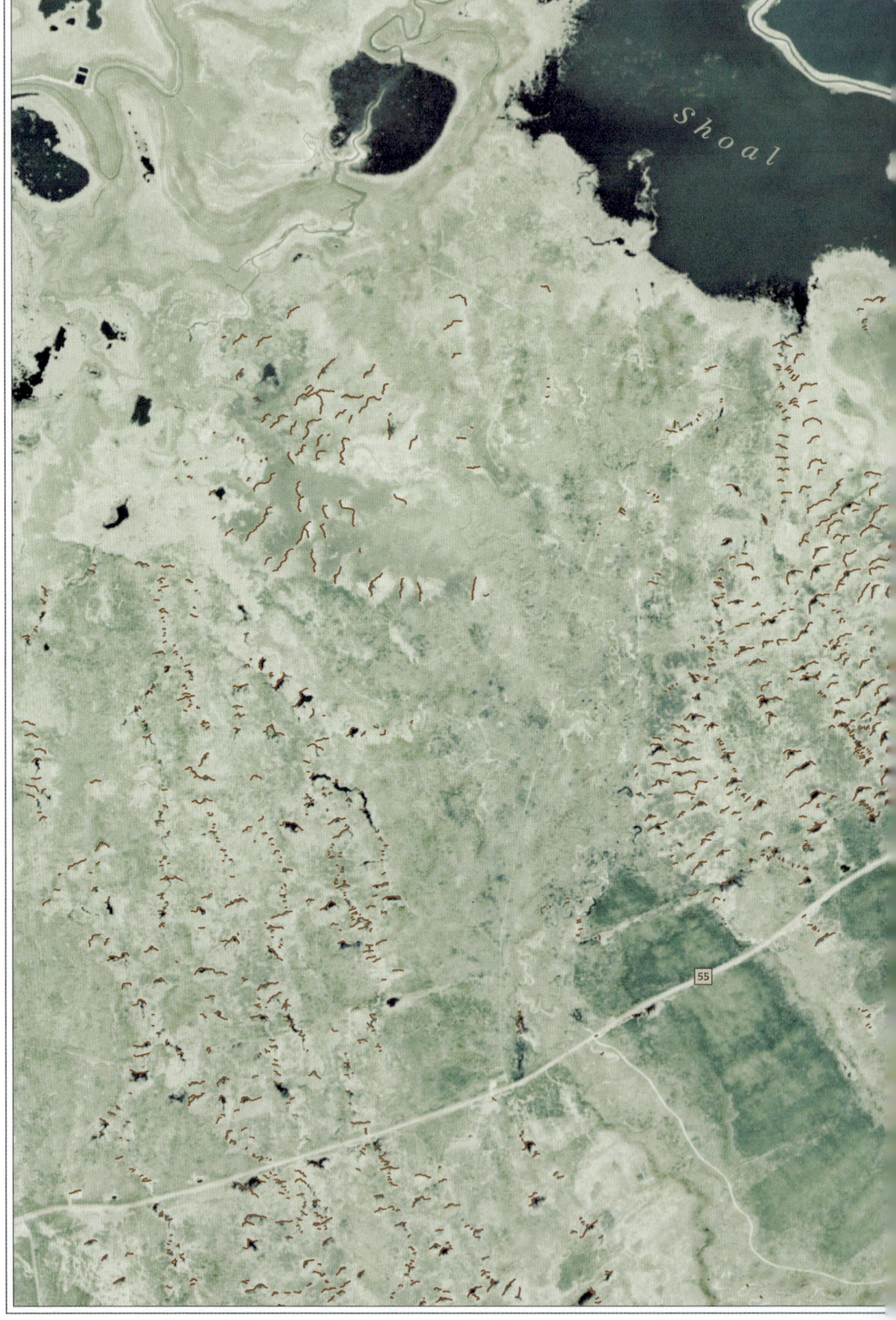
Shoal
55

Lake
PAKWAW LAKE
(Population: 1,090)
More than 50 dams per square mile
55
Beaver dam
Beaver belt
Pakwaw Lake
Beaver dams manually traced from high-resolution satellite imagery, 2024
0 miles
1

detected. This is generally prompted by natural events, such as a period of unusually heavy rainfall that causes a dam wall to buckle and burst. As fur trappers, they would often exploit this instinct, he said, by intentionally springing a small leak in a wall as a means of luring the beavers out.

In one spot, beavers constructed a canal that led west to east, instead of in the water's normal south-north flow. This completely altered the landscape because it flooded areas that would have otherwise been dry. I asked Jerry why they might do this. He wasn't sure, but speculated that it may be to create a new pond to the east or to support an existing one that's too low. They always have a reason for their engineering, he said, even if we don't understand it.

Beavers also commandeered the town's man-made reservoir, Carl noted. Within two years of its completion, the animals moved in, building a giant lodge smack in the middle. The lodge was later abandoned for unknown reasons, but was then swiftly reclaimed by a different beaver family, who, Carl noted, moved the structure's front door from one side to the other. During our visit, the perimeter canals leading into the woods were freshly trampled and covered in scattered tree fragments—telltale signs that the reservoir was once again a populated place.

After weeks of turmoil and hundreds of miles traveled, I finally felt at ease in Pakwaw Lake, seeing with my own eyes how profoundly other animals can remake a place in their image. Left to their own devices, clever rodents have built and rebuilt this terrain to suit their evolving desires, and the land is richer for it.

Surveying their achievements, I was reminded how admirably self-sufficient beavers are, more so than even the most pioneering of human settlers. It would be rare to find a human family living in complete isolation, because as a species, we generally require the support of a larger community to satisfy our many needs and expectations. But each of the thousands of beaver homesteads in this corner of Saskatchewan is a world unto itself, capable of thriving separate and apart from its many neighbors.

After bidding farewell to my guides at the end of our last day, I decided to make the most of the fading daylight hours by returning solo to one of the ponds closest to the highway. Staking it out at dusk, I sat silently in my Kia, windows down, and listened intently to the heartbeat of the forest. Every passing vehicle slowed to a stop when they saw me idling on the roadside. Most assumed I had spotted a moose—an exciting event even for locals. Others figured maybe I needed a jump start. All gave me a baffled look when I said that, actually, I was watching for beavers.

Around 7 p.m., with the sun hanging low in the sky and birds flying home from a busy day of foraging, the fog rolled in like a privacy shroud. I walked down to the shoreline and camouflaged myself behind a reed wall. The pond, deathly still in the summer-evening air, suddenly rippled as a small, brown face stealthily arose from below the waterline like a submarine. I watched as he floated effortlessly around the perimeter of the pond, surveying his territory.

Splashing quickly below the surface, he disappeared for what seemed like a solid thirty seconds before reemerging near the pond's edge, this time with twigs and mud in his mouth. He then thrust himself out of the water and onto the dam wall, set down his supplies, took a precautionary, panoramic scan around, and started using his front paws to reinforce the exact part of the wall where we'd stood earlier in the day. As he did so, a second beaver emerged on the other side of the pond, cruising around with only the faintest wake behind her—more *of* the water than in it.

2

Urban Flight

Cities

AS DUSK DESCENDED over the live oaks and prickly pear cacti of Texas Hill Country, where the South ends and the Southwest begins, a sense of panic was spreading. It was the mid-1800s, and German settlers were engaged in a territorial dispute with the powerful Native American tribes who had long inhabited this land. Fatigued and jittery, the colonists were alarmed to see what appeared to be a massive smoke column on the horizon, blacking out the summer sun.

The settlers raced on horseback to identify the source and steeled themselves for battle, venturing deep into the hills northeast of San Antonio before arriving at an ominous crack in the earth's surface. This was no ordinary smoke, they soon realized. In fact, what they saw wasn't smoke at all.

Unbeknownst to them, the settlers had stumbled onto one of our planet's most awe-inspiring animal aggregations: a tornado of millions of Mexican free-tailed bats spiraling toward the heavens.

THIS STORY, PASSED down by word of mouth through the generations, is almost certainly a blend of fact and folklore. Texans never met a tall tale they didn't like.

But the colossal bat community this anecdote references requires little by way of exaggeration. Located in what is now known as Bracken Cave, it stands as living proof that, apparently, even animal cities are bigger in Texas.

There is no universal criteria for what defines a city beyond the dictionary definition of an inhabited place of comparatively greater size, population, or importance than a town or village. Generally speaking, it's understood to be an area in which a large number of individuals gather or live in close physical proximity, usually to gain certain community advantages. In practice, cities also often serve as centralized hubs where cultural activities are carried out. (In the UK, for instance, a city's status was once tied to whether or not it had a cathedral.) It's where the action happens, bound by a collective spirit that gives a city its beating heart.

Animal cities are not measured by their concrete, glass, or steel, but by the community itself and the tangible advantages that large-scale cohabitation provides. Safety in numbers, knowledge sharing, services and trade, and environmental changes tied to a population's sizable footprint are just a few possible reasons for coming together en masse rather than going it alone.

Even in the winter low-season, Bracken boasts a year-round population of roughly one hundred thousand bats—large enough to elicit at least mild curiosity from human neighbors. But it's around March that this place truly starts to come alive.

For the past 8,500 to 10,000 years, the community inside the cave has swelled exponentially every spring as up to 10 million free-tails fly 700 miles north from various smaller roosts in and around central Mexico. By August, the bat population will double again to nearly 20 million—the largest congregation of nonhuman mammals on Earth.

While the Bracken bats benefit in many ways from gathering in such large numbers, there is one service that makes this particular cave worth the treacherous journey: its incubator.

NEARLY ALL THE Bracken migrants are female, and they are with child. The males who impregnated them back in Mexico also make the journey north around the same time, following a route that predates—and largely ignores—humanity's national borders. But rather than staying in Bracken with the females and their future offspring, most males prefer to hang out in what researchers refer to as "bachelor colonies" under bridges and in smaller caves around the region. The exurbs of the Greater Bracken Metropolitan Area, so to speak.

As the females pile into the cave by the millions, the mercury rises. Mexican free-tailed bats require very warm environments to be comfortable, and Bracken's geological formation is naturally draftier and cooler than they would otherwise want. But that all changes when they come together. By June, the collective body heat of millions of expecting mothers, each smaller than a human hand, raises the temperature inside the cave to a sweltering 106 degrees—ideal conditions for child-rearing.

The millions of Bracken moms will then each give birth to exactly one pup, who weighs about as much as a dime. (Sounds small, but that's roughly the body-scale equivalent of a human mother delivering a thirty-pound newborn.) The pups are placed inside makeshift creches in the crevices of the cave's enormous nursery alongside millions of other babies, where they will be fed several times a day.

For decades, it was textbook knowledge that these bats nursed indiscriminately, says Gary McCracken, a field biologist at the University of Tennessee, Knoxville. Because millions of mothers simultaneously feeding millions of pups looked so damn chaotic, many scientists simply assumed that the moms must be giving their milk away to the first hungry mouth they find, leading the strongest pups to prevail and the weakest to starve.

McCracken, ever skeptical of the idea that human limitations can or should be applied to other species, set about disproving this fallacy. Through years of careful observation and genetic analysis, he conclusively showed that female free-tailed bats actually vocalize unique directive social calls, which, when combined with their powerful sense of smell, enable moms to precisely identify and feed their baby among the hordes.

After feeding, the mothers return to their corner of the cave and crowd together cheek by jowl, effectively turning their bodies into what one expert described as "a radiant heater."

By July, this bat metropolis is bursting at the seams. Millions of hungry pups grow restless, and food demand reaches its zenith. The result is a nightly commute of epic proportions, as legions of nocturnal hunters, young and old, flood out of the city and into the surrounding fields—an event known simply as "the emergence."

At its peak, Bracken's population is so dense that it takes over three and a half hours for every bat to hurtle out of the cave's bottleneck entrance, creating a plume of tiny flapping wings that resembles a snake extending into the evening sky. Soaring up to 10,000 feet, the moms and pups fan outward, traveling between 60 and 100 miles a night and collectively consuming more than 150 tons of moths, beetles, and other insects, before returning to their hideaway home just before daybreak.

AS I DROVE past the miles of insurance billboards, car dealerships, fast-food restaurants, and gentlemen's clubs that dot San Antonio's suburbs, the visuals were considerably less impressive.

Only after passing Randolph Air Force Base and the smallish town of Schertz did the rolling hills of Comal County start to return to their more natural state of parched golden meadows, spindly trees, and limestone. With the windows down, DIY bed set up in the back seat, and radio speakers blaring '70s Americana, I hit my stride on the open road. The background buzz of chirping cicadas was a reassuring sign that I was getting closer.

It was hot out. Very hot. Highway-looks-fuzzy hot.

The Hill Country was in the thick of a record-breaking heat dome that would ultimately fuel more than forty consecutive days of triple-digit temperatures. At local summer camps, outdoor play had been strictly limited to fifteen minutes to keep kids from keeling over on the broiling pavement. I passed through the security gate of the Bracken Cave Preserve around dinnertime, as hungry hawks circled ominously overhead.

A dusty gravel road guides visitors to the center of the preserve, which Bat Conservation International (BCI) purchased from a private landowner in 1992. Originally just a five-acre plot, conservationists have since expanded its footprint through a number of adjacent acquisitions. They also successfully thwarted a proposed housing development that was slated to build 3,500 McMansions along this entrance route. Through that victory and others, BCI's preserve now spans 1,500 acres, creating a natural buffer of open grassland and dark skies between Bracken and humanity's urban sprawl. Hailing the 2014 agreement that kept encroaching developers at bay, San Antonio City Councilman Ron Nirenberg—who later served as mayor—praised the preservation of what he described as "the glory" of the

Texas Hill Country, adding that people who aren't moved by sights like the Bracken emergence should probably "check [their] pulse."

As I took my seat on a small viewing platform that overlooks the sinkhole entrance, volunteer docent Karla was feeling optimistic. "Y'all are here on a good night," she said. "I know it's hot, but it's always best to come when it's hottest. If you can suffer through it, you'll be greatly rewarded."

Besides the ambient heat, one of the first things visitors notice when approaching the cave entrance is the acrid scent of guano—bat poop. It smells like a smoldering dumpster. Once guano hits the ground, it can remain there for thousands of years, sometimes halting future stalagmite growth and changing the cave's landscape.

Prized by farmers as a prolific fertilizer, it reigned as Texas's largest mineral export until the oil boom at the start of the twentieth century. But it also has a more explosive application. Guano contains potassium nitrate, which, when mixed with sulfur and charcoal, can produce gunpowder.

That alternative usage became all the more relevant in the final years of the Civil War as Union blockades successfully disrupted Southern shipping routes, leading to severe munitions shortages. Suddenly, what was then known as Cibolo Cave took on new strategic importance. Confederate soldiers started mining the cave in the winter months to haul out guano, which was then transported to nearby New Braunfels for nitrate extraction. A Houston-bound railway station was later built in the nearby town of Bracken, leading to the cave's updated moniker.

The soldiers were wise to limit their digging to the off-peak season, not just because of the cooler temperatures, but also to avoid incurring the wrath of flesh-eating beetles who inhabit the guano during the summertime. The beetles are so voracious that if a baby bat accidentally falls from the cave ceiling, the pup will be reduced to a cleaned skeleton within hours.

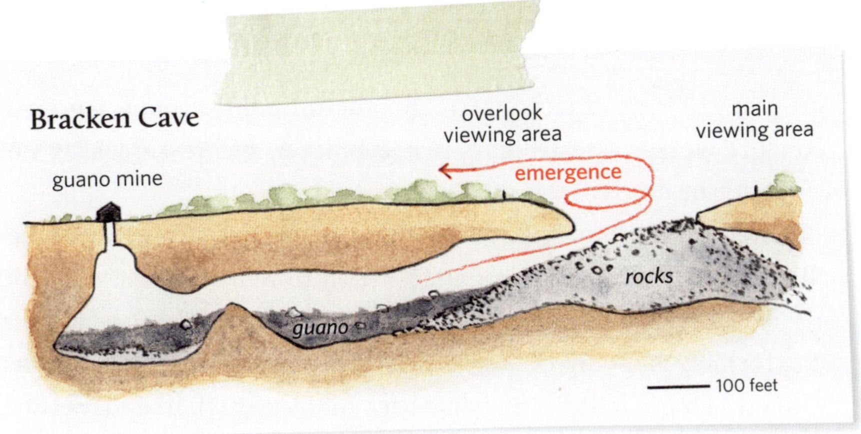

In their frenzy, the Confederacy is believed to have extracted up to 2,500 pounds of guano a day. But even at that accelerated pace, there was never much worry about depleting the supply. In many caves, guano accumulation is no more than a few feet thick. But in Bracken, with its millennia-long history of bat residency, depths can exceed six stories.

AS THE SUN began to sink on the horizon, momentum was picking up. Peering into the crack in the earth through the lens of a thermal scope, I saw a white-hot vortex swirling in the dark—the beginnings of what BCI refers to as the "Batnado."

Unlike birds, most bats cannot just flap their wings and take off from the ground. They have to build up speed and incrementally spiral upward, or fall from above and catch the wind like a paraglider. Once airborne, they can fly at up to 99 miles per hour—the fastest horizontal speed of any animal*—but they need a lot of room to

* The overall speed record for the animal kingdom goes to the peregrine falcon, whose vertical dives have been estimated to reach an astonishing 242 miles per hour. It's easy to see why they make mother bats so skittish.

get going. That's especially true for the millions of pups, who will conduct weeks of test flights inside Bracken before careening out through the cave's exit for the first time.

Bracken is suitably large for the bats' particular needs. Three-dimensional lidar scans have pegged it at roughly 124 feet wide and 117 feet tall—from guano to ceiling—meaning Lady Liberty could comfortably stand up or lie down inside with a few feet to spare. It's also very long, extending 650 feet back into the earth like a lava tube.

I was joined at this viewing by BCI members and supporters, many of whom had flown in from around the world just to observe the spectacle. "It's a visual experience, but it's also an auditory one," Karla said. "If we're very, very quiet and you hold your hands up to your ears, you'll hear what sounds like a faraway waterfall. That's the sound of those fifteen to twenty million bats in there, flapping their wings." She's right—that's exactly what it sounds like.

We don't yet know what triggers the moment that the bats choose to exit the cave, but there is clearly a complex coordination process taking place. The docents have noticed that during periods of drought or extreme heat, the bats tend to start their nightly insect foraging earlier, perhaps anticipating that they may need to travel to more distant corn and cotton fields to get their fill. And certainly once the fledgling pups start accompanying their moms on the nightly hunt, additional precautions are taken to ensure that the coast is clear.

Dr. Winifred Frick, BCI's chief scientist, says that the perceived security of sharing a city with millions of other concerned moms is part of Bracken's enduring appeal. When bats organize themselves in such massive numbers, it can overwhelm the senses of hawks and other predators who would love nothing more than to turn one of these evening commuters into dinner. That heightened sense of safety entices females to come back year after year. Some will spend up to fifteen consecutive summers in Bracken, and females born in the cave are very likely to return when they're ready to reproduce.

The good word also appears to spread beyond current residents, Frick says. "The fact that so many other bats are clearly thriving in this 'city' is no doubt hugely appealing to potential immigrants. The community's sheer size is proof in some sense that it's a safe place to live, compared with a newer, less proven location. That's tremendously reassuring to someone who's about to become a mom." As they enter old age, many nonreproductive females will depart Bracken for the last time, joining the bachelors in their exurban colonies as a generation of new moms moves in.

IT'S RISKY TO start any sentence with "Bats do this . . . ," because bats are an enormously diverse umbrella category, spanning more than 1,400 species. One out of every five mammals is some kind of bat, and the differences between bat species are as wide as those between "a Siberian tiger and a sea otter," says Merlin Tuttle, who founded BCI before starting a new venture, Merlin Tuttle's Bat Conservation. Depending on the species, a bat's wingspan can stretch from a mere six inches to nearly six feet. Most gather in communities, but some are solitary. Some eat fruit, but others prefer fish, cattle blood, frogs, or something else entirely.

That said, as a general rule, the larger an animal community is, the more information is being passed around, and the better its residents can collectively sense the nuances of the landscape. The increased swirling of knowledge around a community can, in turn, contribute to a sort of collective intelligence.

Some fruit bats, for instance, maintain a mental map of all the trees that could potentially fruit, but they don't always know which ones will bloom next. The speculative knowledge they gather while sussing out the local options is then shared by scent with their close contacts as part of strict social groups. We don't know how tight-lipped or active the Bracken bats are about sharing their learnings, but Gary McCracken, the researcher who proved that moms feed their own pups, says that "there's definitely eavesdropping going on."

Similar knowledge-sharing dynamics have been observed in other non-bat species throughout the animal kingdom.

Professor Emily Shepard at the Swansea Lab for Animal Movement told me how she would sometimes visit North Atlantic islands that had thousands of seabirds clustered in one area, while a nearby cliff that looked, by all appearances, to be equally favorable had none. To try to make sense of why the birds chose to position their temporary cities in certain places, she started looking at the differences between the various cliff options. Topographically, many seemed of roughly equal height and afforded similar protection from predators, so why not spread out a bit more? The answer, she said, was blowin' in the wind.

Wind drives many of the most important decisions for seabirds. Some stake out sheltered areas to protect their chicks from an icy chill, while others congregate in exposed places where strong gusts crash into the landscape, which may help them more easily take off.

Upon discovering this connection, Shepard and her team started applying this methodology to other islands where they had not previously observed bird communities. By modeling an island's wind flows and slope angles, her team would attempt to "visualize the invisible" by making educated guesses about where seabird communities would be most likely to establish themselves. In more than 70 percent of cases, their predictions proved spot-on.

In such situations, Shepard says, the benefits of community clustering clearly go beyond safety in numbers. Knowledge about where the best cushions of air can be found, or potential insights on prime feeding spots, can be extremely valuable commodities to exchange. When birds come together en masse, more information circulates, enabling the entire group to make educated decisions.

Similar calculations are no doubt made by many other birds around the world, she says. Andean condors—who can weigh more than thirty pounds—place a particularly high value on finding winds strong enough to keep their bulbous bodies aloft, knowledge of which Shepard believes they likely exchange. GPS tracking has shown that vultures definitely share knowledge of food locations with their peers, congregating in cities of more than one hundred residents, so it would make sense that knowledge of favorable wind conditions would also be part of their conversations. "They're responding to the weather," Shepard says. "They might even be forecasting it—we don't know."

BEYOND KNOWLEDGE SHARING, there are other factors that can make life in an animal city appealing. More than one hundred thousand giant tortoises live on a single coral atoll in the Seychelles archipelago, which provides them with a vast pool of potential mates. The males mate with as many partners as possible, but the females have such a wide array of options that they can afford to be selective. If you're a giant tortoise seeking a mate, you want to be in a dense city, not stuck in a small town.

The same could be said for the 2.5 million flamingos who spend at least part of their year in Tanzania's Lake Natron. The birds have learned that their chicks have much better odds of survival in a large community of like-minded peers, where many eyes provide greater protection from predators.

In North America, perhaps the most famous example of large-scale animal communities are the iconic prairie dog "towns" of the

Great Plains. Living at densities of five to thirty-five residents per acre, prairie dog towns have clearly defined nursery areas, food storage chambers, and lookout spots for keeping an eye on approaching threats. The areas they span are often large enough that we can now identify them from satellite imagery, similar to how Jean Thie spots beaver dams. While human interference over the past century has contributed to a significant decline in prairie dog population numbers, historical records by renowned naturalist Vernon Bailey show that as recently as 1900, a single prairie dog town—or perhaps more accurately, city—in North Central Texas was home to an estimated 400 million residents and spanned 25,000 square miles—about forty times the size of modern-day Houston.

Large-scale animal communities are, however, not without their drawbacks. If a group size grows too much too fast, it can strain finite natural resources like food and water, potentially leading to an exodus of longtime residents or even a full-scale societal collapse. In the animal kingdom, there is no such thing as too big to fail.

Dense animal cities are also more susceptible to mass extermination than a dispersed network would be, says Merlin Tuttle. He recounted an example of a protected cave in Florida where he watched a bat community grow from a couple thousand residents to more than 4 million. But then, the cave was directly hit by a hundred-year hurricane, causing most of the bats to drown. Twenty years later, long after the floodwaters subsided, Tuttle says, very few bats use that cave. He believes this is a result of horror stories and dire warnings being publicly communicated by the storm's remaining survivors. "They don't hand down knowledge through books. They hand it down bat to bat." If you wipe out an entire community, the culture and knowledge their ancestors shared goes with it.

BACK AT BRACKEN, the small human crowd gathered outside the entrance was sweating with anticipation. Just inside the lip of the cave, the bats were coiled around a slowly escalating vortex as the Batnado picked up speed.

Then, at exactly 8:18 p.m., they made their move, crossing the Rubicon into the open air and coasting upward on the sunset winds. Cunning coachwhip snakes lurking in the brush excitedly popped up, having waited all day to pluck an unsuspecting pup out of thin air.

Bursting en masse into the breeze, the bats quickly ascended, flowing like a winged river through the golden-hour sky. Upon reaching a cruising altitude of several thousand feet, the river turned southeast, splintering into a series of smaller streams destined for various agricultural fields.

We don't yet understand what factors determine how these smaller streams are organized. Long-distance GPS tracking of bats is still a nascent field, in part because they travel at night, rendering the latest generation of solar trackers powerless. Some researchers speculate that their sky groupings are based on crowds following key individuals who are perceived to have useful knowledge about where the good eats can be found—a sensible assumption, given how much knowledge is no doubt floating around within the community.

Fran Hutchins, director of Bracken Cave Preserve, has an alternative theory. Having witnessed more Bracken emergences than almost anybody, he's come to believe that the smaller sky streams actually reflect bats flying alongside family friends from their days back in Mexico; a tantalizing—if admittedly unproven—idea that suggests the bats maintain what are effectively ethnic enclaves within Bracken's urban landscape.

BATS ONLY BEYOND
THIS POINT

Some experts, including Martin Wikelski from ICARUS, are skeptical. “I don’t see any reason why they would keep the same alliances. I could easily imagine the opposite, given that other bats might have better information.” With so many choices, “they don’t have to follow their friends.”

Others, like Tuttle, were careful to leave the door open, having seen bats shatter human expectations and assumptions many times before. He personally spent decades putting colored bands on more than forty-two thousand gray bats in an attempt to better understand where they go and how they organize themselves. Over a twenty-year stretch, he periodically captured and recaptured the banded bats in various locations. During that time, he was repeatedly astonished to find the same small bat groupings clustered together year after year, sometimes 100 miles away from where he first banded them. “We now know from fairly recent research that bats have [a] social order strikingly similar to higher primates’,” he said. “They help each other. They share information.” So ethnic enclaves inside Bracken? “None of this is unbelievable.”

Leaning back in his chair, Tuttle recalled how turbulent summer-evening storms sometimes created unique challenges for the Bracken emergence. From the mouth of the cave, he would watch as the bats began their nightly foraging earlier than usual, flying thousands of feet into the sky, before nose-diving back into the cave after twenty to thirty minutes “like little jets.” Then, a few minutes later, a second, much larger emergence would begin.

“I see no earthly reason why those bats would have gone out early and taken a risk getting caught by a hawk if they weren’t accomplishing some important purpose, which is probably communicating about where the best tailwinds are to ride on amid the turbulence. Information transfers about where to go, how to get there, all sorts of things. Bats are good communicators. I don’t know if they stick together all the way from Mexico, but are we really going to say it's not possible? I certainly wouldn't."

By midsummer, when the moms and pups take flight together, the emergence continues long past sunset, with the last bat setting sail sometime between 11 p.m. and midnight.

As I watched this mesmerizing scene play out on a Doppler radar app, one of the docents held up his phone to show everyone the live feed. The nearby New Braunfels radar station, which continuously scans the horizon for approaching thunderheads, soon picked up a massive blob larger than metropolitan San Antonio. Weather apps generally modify their default settings to remove such scenes from view, the docent said. Otherwise, even on a clear night, people watching the radar feed would break out their umbrellas, believing they were about to get caught in a torrential downpour.

RETURNING TO BRACKEN a few days later, I made special arrangements with Fran to go past the BATS ONLY sign at the edge of the public viewing platform and walk down to the mouth of the cave itself. In summertime, this is the closest that any human can reasonably get to Bracken's heart without risking violent illness, injury, or death, given the toxic levels of ammonia inside. At Fran's suggestion, I wore a thick N95 mask, despite the stifling heat.

From this closer vantage point, I had to keep my wits about me, recalling the skulking snakes I'd seen on my previous visit.

I approached the bone-dry edge of the cave mouth to see the subtle undulations of the emerging Batnado on full display. Bracken is almost never devoid of activity, Fran said. This is a city that never sleeps, though the bats take turns "catnapping" on and off throughout the day. As the small cyclone whipped closer, its intensity grew.

With every rotation, thousands more bats leapt from their niches on the cave wall to join in, like a giant spool of black cotton candy.

The suspense was palpable as I looked for any visible clues that the bats were about to gun it for the exit. Just then, a few feet to my left, a large coachwhip hungrily slithered down the hillside, strategically positioning himself at the entrance.

If there was a decisive moment that prompted the bats to make their move, it was beyond my perception, but just before 8 p.m., they collectively shimmied up to the exit and burst out into the open air.

Crouched as close as I could get to the ground, I soon had tens of thousands of seemingly identical-looking bats barreling around just inches above me, generating an astonishing amount of wind. A few gently clonked me on the side of the head before rebounding and continuing onward.

As they picked up speed and flooded out with growing intensity, my eyes strained to keep pace while my poor brain all but gave up trying to interpret what it was taking in. My heart also raced, knowing that by hunkering below the eye of the hurricane, my body had in that moment become part of their landscape. Just a passive object to shove the Texas humidity onto as the bats collectively thrust upward, creating their own weather and forging a path of least resistance for those in their wake.

AFTER WAKING UP the next morning in my car—a bachelor colony of one, ambiguously straddling big-box parking lots—I made my way to the locker room of a nearby public pool and scrubbed off the caked-on layers of dust, dirt, and worse.

Around the same moment that Bracken's bats were making their move the previous night, other communities around Central Texas were also taking to the skies. Perhaps the most famous example lies 60-odd miles northeast, as the bat flies, under Austin's Congress Avenue Bridge.

When city engineers began renovations on Congress Avenue in 1980, it was determined that the steel rebar that reinforced the bridge's concrete superstructure was overstressed, and the cantilevered overhangs were deteriorating, meaning they effectively had to take the whole top off. As part of the reconstruction, new concrete box beams were also installed on its underside, spaced between half an inch and two inches apart. Mark Bloschock, then a mid-twenties inspector on the bridge, worked nearly every day for two years to see the construction through to completion and was justifiably proud of how it turned out.

Then, the first call came in. A frightened jogger on the riverside hike-and-bike trail had spotted what appeared to be a small group of bats clinging to the underside of the structure. Local authorities were called in to make a sanitary sweep.

Soon more started arriving. And more after that. As summer dawned, it became clear that the narrow gaps between each of the support beams Bloschock and his colleagues had painstakingly installed were just the right dimensions for free-tailed bats: wide enough to slip their bodies in and out of, but skinny enough to limit their open-air exposure. The concrete also conducts heat, creating sun-scorched micro-caves with an ideal ambient temperature of up to 110 degrees. Add to that a strategic location over a stretch of the Colorado River known as Lady Bird Lake ("Town Lake" among Austinites), which kept snakes—and barhopping tech bros—at a distance. This, they decided, was the ideal place to raise kids.

Within two years, more than 1.5 million residents had arrived—the largest bat community within a (human) urban area anywhere in North America.

Austin city leaders were alarmed. Horror stories started circulating, falsely claiming that the migrants were spreading rabies and sucking people's blood, with an alleged preference for women and children. Petitions were circulated demanding the animals' immediate extermination.

Concerned by the simmering human-bat conflict he read about in the papers, Tuttle decided to relocate Bat Conservation International's global headquarters from Milwaukee to Austin. When he arrived, the new bridge dwellers were so universally despised that the mere notion of a bat protection group moving to Texas earned him an ignominious Bum Steer Award from *Texas Monthly* magazine—a title awarded annually to "the dopes, villains, and terrible ideas that bedeviled our beloved state over the past twelve months."

During one of his earliest visits to Bracken, long before BCI bought the land around it, Tuttle remembers watching a group of good ol' boys drinking beers and watching the emergence. As he got within earshot of them, he overheard them casually speculating about how much fun it would be to throw sticks of dynamite into the cave and watch all the bats frantically try to escape.

Around this same time, Bloschock was strolling down "the Drag," which runs along the western edge of the University of Texas campus, and wandered into a local provisions store. On the shelf was one of Tuttle's books, titled *America's Neighborhood Bats*. With the Congress Avenue fury top of mind, he picked up a copy and read it cover to cover. He then reached out to Tuttle to ask if they could meet up for lunch.

"Supporters are calling me to complain that the Texas Department of Transportation is burning bats alive with blowtorches," Tuttle told him. (An accusation that TxDOT officials dispute.) Tuttle also sought to dispel the myths that had been circulating one by one. No, these bats don't suck blood. In fact, they eat bugs, which is advantageous for local farmers. They aren't any more likely to spread rabies than other mammals (less, actually). And no, they don't attack people. They want nothing to do with us.

Bloschock listened intently. "Tuttle was smart, because he knew that engineers love to get double or triple duty out of anything we do." If this bridge could both serve the people of Austin and also provide safe refuge for wildlife without compromising the integrity of the structure, "well, heck, that's just more bang for your buck."

A working relationship soon blossomed as Tuttle and Bloschock tried to jointly turn the tide within the city leadership. Bloschock started inviting Tuttle to be a featured speaker at city planning meetings and state engineering conferences to help officials see this as an opportunity, not a crisis.

The bats' presence was a blessing in disguise, Tuttle said. The riverfront area where the free-tails had taken up residence was in the process of being reimagined by city planners, who envisioned a newly gentrified hotspot in downtown Austin. The bats, he said, could be the neighborhood folk heroes. Winged icons of a city on the rise.

The PR campaign worked. Not only did resistance slowly start to dissipate, but a multimillion-dollar cottage industry soon sprang up to celebrate the bats as an integral part of Austin's much-hyped weirdness. Kayak and electric-boat tours started taking paying customers under the bridge to observe the nightly emergence. Kiosks along the surrounding walking paths directed tourists to the best viewing spots. The nearby Hyatt Regency hotel opened the riverfront Bat Bar, and bat-themed street art began appearing all over the city.

In 2021, as a sign of shifting perceptions, Tuttle's infamous Bum Steer Award was officially rescinded by *Texas Monthly*, and the mayor declared August 26 "Merlin Tuttle Day" in the city of Austin.

Curious to witness this bat fandom with my own eyes, I drove the ninety minutes up from Bracken to Austin on a Saturday afternoon to join a few dozen other passengers on a $13 sunset tour. The hokey jaunt was filled with cringy tourist trappings, spearheaded by our boat guide, who identified himself only as "Captain Han Solo." A recent UT graduate, Solo spent most of our time on the water cracking jokes at the expense of rival colleges and cities.

After taking a tedious detour upriver to point out the various Silicon Valley companies that have come to dominate the Austin skyline, the captain turned us back toward the bridge. To my surprise, during the brief time we were spinning around looking at skyscrapers, Congress Avenue had filled up end to end with gawking spectators, all anxiously awaiting the nightly flight. The bridge was clearly *the* place to see and be seen on a sweltering Saturday night. The river was also replete with kayak-bound tour guides, who shined red lights up at various subsections of the bridge, which they identified as "Neighborhood 1," "Neighborhood 2," and so on. (Bats can't see red as well as other colors, Solo said, so the choice of lighting is intentional to avoid simulating daylight and delaying the bats' flight.)

Ensconced in their concrete canyons, the bats were engaged in a symphony of soothing chirps known as "colony chatter."

Captain Solo warned us before departure that the bats had not been especially cooperative with his itinerary lately, coming out later and more sporadically than his customers often expect. Especially this time of year, he said, the bridge is bursting with nervous new moms, who are keenly aware that if they get eaten by a hawk, their baby will starve. Asked by a passenger how many bats we were guaranteed to see, Solo—only half jokingly—said, "At least one."

But then, just moments before sunset, we caught a break. "Oh, look—here they come!" Solo giddily heralded. "Wow, y'all are lucky. This is the best timing we've had in a month and a half." The evening air was soon swarmed by thousands of bats as they dropped from the bridge and caught the breeze.

Apparently sensing that this was his moment to shine, one of the male passengers on board took the opportunity to awkwardly stand up and make his own announcement. "Ummm, Captain Han Solo, I have a question. You were talking about how . . . like . . . the female bats get with the males, and it got me thinking . . ." He then carefully lowered himself onto one knee and proposed to his girlfriend right there under the emergence.

Overwhelmed, she enthusiastically accepted, and the crowd went wild.

MOVED BY THE transformation in public sentiment he'd witnessed, Mark Bloschock would go on to spend the next three decades reno-

vating hundreds of bridges, overpasses, and culverts across Texas, using Congress Avenue's bat-friendly crevices as a template. After his retirement from the agency, TxDOT continued to build out his legacy statewide.

In the sprawling suburb of Round Rock, I wandered around under the McNeil Overpass, which at first glance appears to be nothing but a concrete slab adjacent to an abandoned auto-parts store. Most people driving on or under it have no idea that it's now home to an estimated 2 million bats. Even in the middle of the day, the colony chatter is astonishingly loud, when not drowned out by the tailpipe farts of passing vehicles.

As fall approaches, free-tails across the Hill Country, from the McNeil moms to the swingin' bachelors who have spent the past six months chilling in various anonymous caves and roosts, make final preparations for their migration south of the border. In anticipation of departure, the bats will tap into an invisible network of staging grounds—rendezvous points where they can reliably huddle together before the arduous journey.

Congress Avenue has become one such hub, according to BCI signage along the riverfront. Decades after its bat-friendly infrastructure created an accidental haven, the bridge appears to have morphed into a kind of way station. A historic landmark on the bats' mental map.

By the end of summer, many bats who have spent months living away from the hustle and bustle will make a pilgrimage of sorts to downtown Austin, causing Congress Avenue's population to swell to its highest levels of the year.

Only after slumbering together at the tail end of the season will the Austin bats begin to vacate. As the cold weather creeps in from the horizon, they become a city on the move.

BEFORE BIDDING FAREWELL to the Hill Country, I had one final stop to make, at an unassuming bungalow that doubles as the headquarters of Austin Bat Refuge.

Part shelter, part medical center, the nonprofit organization is run by a charming couple named Lee and Dianne. Dianne used to be an education manager at BCI, while Lee's primary profession is as a carpenter and self-described amateur bat specialist. "The word *amateur* comes from the root word *amator*," he noted, "which means 'lover of what we do.'"

As I recounted my experience at Bracken, Lee was delighted by Fran's theory about central Mexican ethnic enclaves within the cave. A bat city populated not by Chinatowns and Koreatowns but by, in Lee's words, "Jalisco-towns!"

He added that there's reason to believe Mexican free-tailed bats could have regional dialects or accents—something already proved among some other kinds of bats, who vocalize "words" differently depending on where they were raised. "I'll bet they have that 'East Texas rust vs. West Texas dust' contrast like we do."

In Lee and Dianne's backyard, I entered a two-story aviary structure filled with dozens of tiny bats in various stages of rest and recovery. One of them, named Penelope after a pin was surgically implanted in her right wing by a sympathetic veterinarian, had a fresh cast on. "Very small surgical tools," Lee said.

All the bats in their care—fifty-two at the time of my visit, not including long-term guests—were either personally rescued by Lee and Dianne after becoming lost or injured or brought to the refuge by a Good Samaritan. As a matter of policy, they never turn away an animal in need, and their goal is to release 100 percent of the bats back into the wild.

At times, this generosity has pushed them to the limit. The refuge went into triage mode during the so-called Great Texas Freeze of 2021, when a polar vortex swept through the region, bursting pipes and causing weeklong power outages. Their phones began ringing off the hook as thousands of frigid bats, who had—perhaps unwisely—decided to stick it out in the Hill Country for the winter, started falling from bridges and overpasses across the state.

As Dianne told this story, Lee's cell phone was dinging like crazy. At that very moment, he was texting with a local twentysomething student who had found a tiny red bat clinging to her apartment window. When the bat didn't leave by the next morning, she reached out for guidance. Being a carpenter, Lee walked her through the process of unscrewing and detaching her window frame from the wall so that she could reach the bat and hopefully put them into a box for safe transit.

"She did it!" he hollered triumphantly. "I'm impressed."

"So she's on the way here?" Dianne asked.

"Yup!"

About thirty minutes later, the woman and her friend walked in, carrying a tiny female in a shoebox. Dianne pulled the bat out and began hand-feeding her with an eyedropper before escorting her to the backyard flight cage. Seemingly satisfied by her good deed, the student promptly took videos of the rejuvenated bat taking flight for her Instagram. "Don't forget to tag Austin Bat Refuge!" Lee helpfully nudged.

Between their various rescue operations in and around the Hill Country, Lee and Dianne may have the most comprehensive firsthand knowledge of where bats congregate of any two Texans in history, though they said they don't maintain any formal lists of confirmed roosts.

"That's a shame," I remarked to Dianne, imagining that such

a resource might increase the odds that their habitats could be protected. From even the brief glimpse I had gotten into the Doppler radar livestream at Bracken, it was clear that there were well-established bat cities and staging grounds all over the region, often hiding in plain sight.

"Well, I don't know if you're familiar with Lee's research that he's been doing for eleven or twelve years. He's done Doppler radar research every single day for over a decade monitoring the emergences," she said as I raised an eyebrow. "He has ninety different points that he gets from Doppler radar data. He probably has more data about bat emergences in Central Texas than anybody on Earth. I'm sure Lee would love to show you what he's got.

"Hey, Lee! Ryan's interested in your Doppler data," she shouted.

Lee grinned ear to ear. "Come with me."

Seated in his home office, Lee explained that several years back he had registered for a special training course offered at the National Oceanic and Atmospheric Administration (NOAA)'s National Severe Storms Laboratory in Norman, Oklahoma. The course was designed for US Air Force pilots and other military professionals who need to monitor Doppler weather information in real time, but Lee wanted to take things one step further.

Using raw data from weather stations in San Angelo, Granger, New Braunfels, and Brackettville—the four corners of what he refers to as the "Significant Bat Area of Central Texas"—NOAA agreed to build Lee a calculus-based algorithm that can translate the on-screen blobs into an estimate of how many individuals are in the air. "It's crazy complicated," he said.

As he showed me the previous night's download, he zoomed out beyond Austin's city limits to point to dozens of other emergences. "That's Bracken there. That's Frio Bat Cave. That's a new place I'm trying to find out," he said, lingering over an emergence in an area far away from any human roads.

"An unexplored cave, you think?" I asked.

"Probably. Nobody else knows about it because nobody's looking—except us!"

The sources were carefully labeled based on Lee's sleuthing. "Chiroptorium," "Old Tunnel," "Randolph Air Force Base"—the list went on and on. Highway overpasses, rural caves, construction sites. Bats make themselves at home in any place that fits their needs and desires, he said. And night after night, for more than a decade, he had downloaded massive data sets driven solely by his own curiosity.

He zoomed back out to the full view and paused as we sat in silent awe, reflecting on all that the radar revealed. As impressive as Bracken is, with its 100 tons of bats screeching nightly through a narrow cave mouth, even that overwhelming sight does not capture the true scale of what's going on.

There are cities within cities. A seemingly infinite blur of Hill Country landmarks, transit networks, and subcommunities that we are only just beginning to understand. An opaque megalopolis of epic proportions, cloaked in the shadow of nightfall.

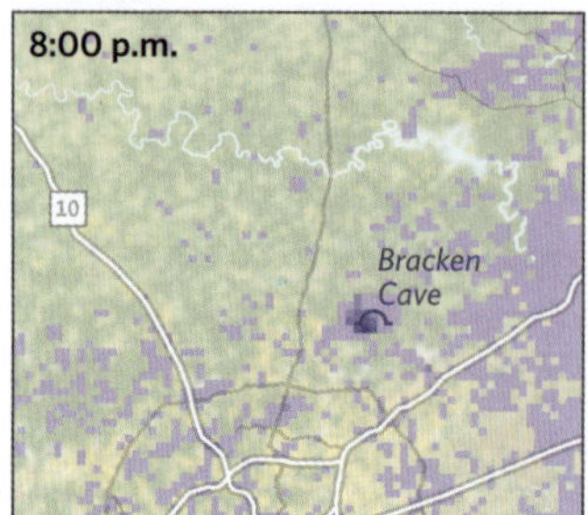

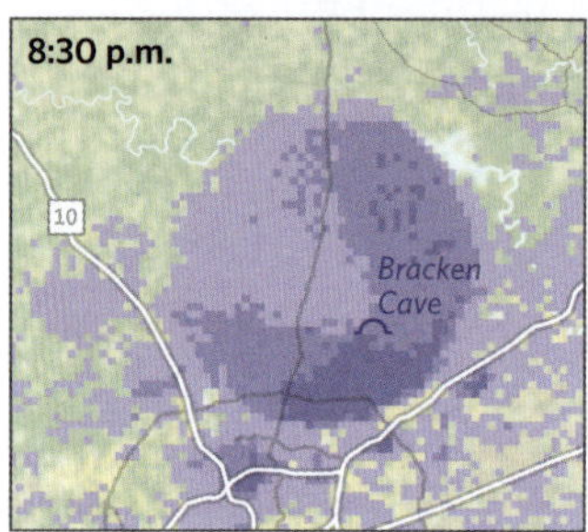

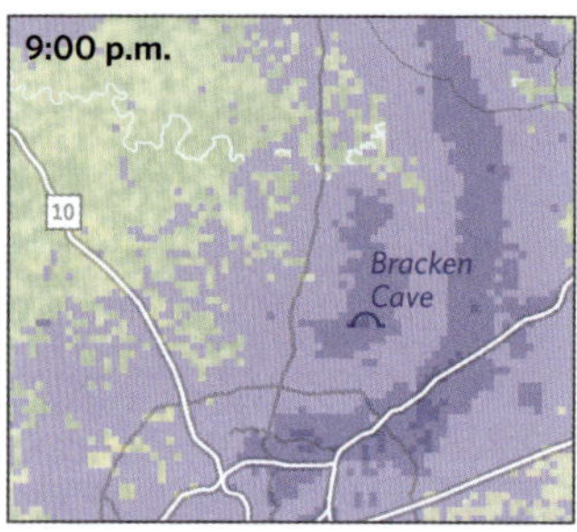

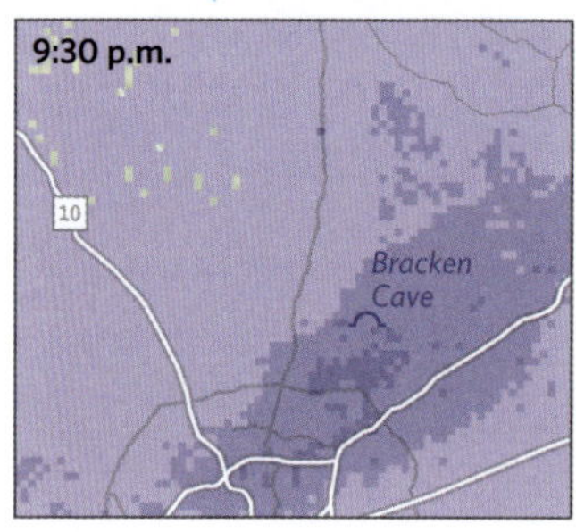

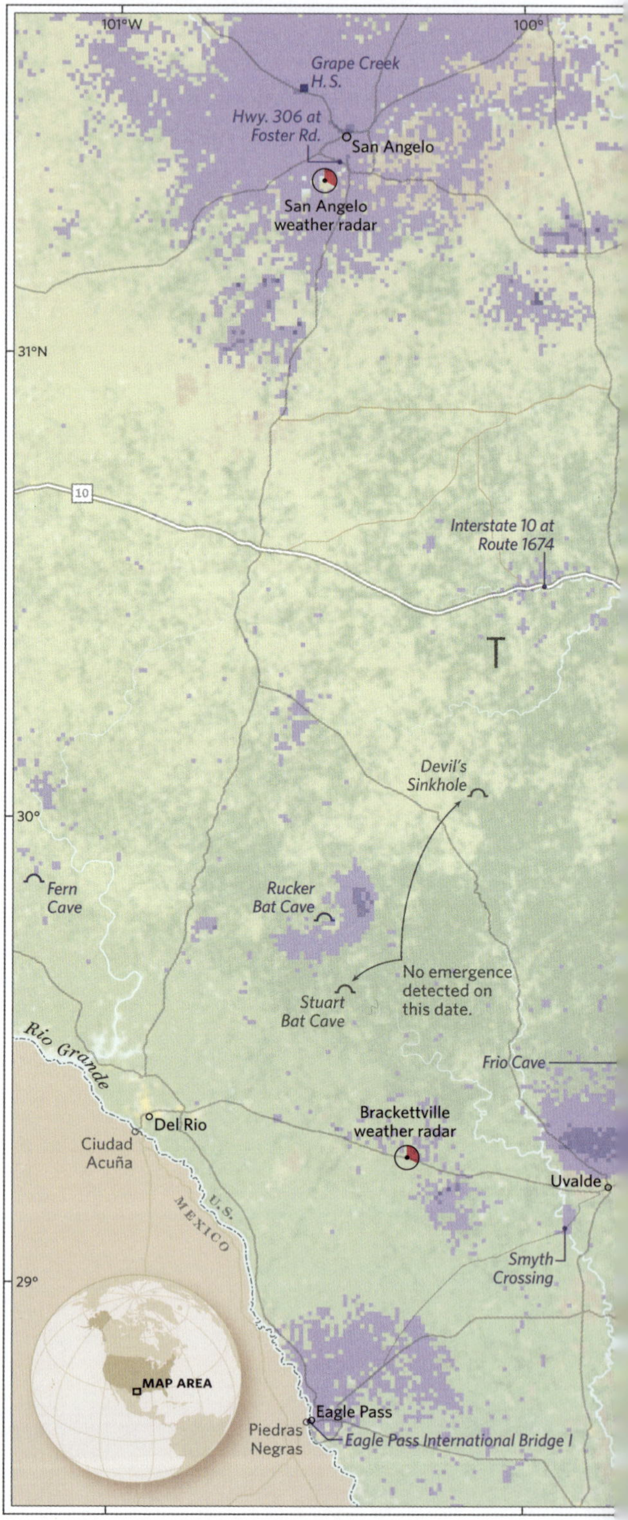

Bat emergences detected by weather radar
July 16, 2023, 8:00–9:30 p.m.

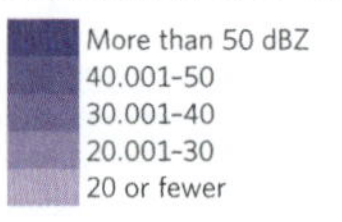

0 miles 25

101°W
100°
31°N
30°
29°
Grape Creek H.S.
Hwy. 306 at Foster Rd.
San Angelo
San Angelo weather radar
10
Interstate 10 at Route 1674
T
Devil's Sinkhole
Fern Cave
Rucker Bat Cave
Stuart Bat Cave
No emergence detected on this date.
Rio Grande
Frio Cave
Del Rio
Ciudad Acuña
Brackettville weather radar
Uvalde
U.S.
MEXICO
Smyth Crossing
MAP AREA
Eagle Pass
Piedras Negras
Eagle Pass International Bridge I

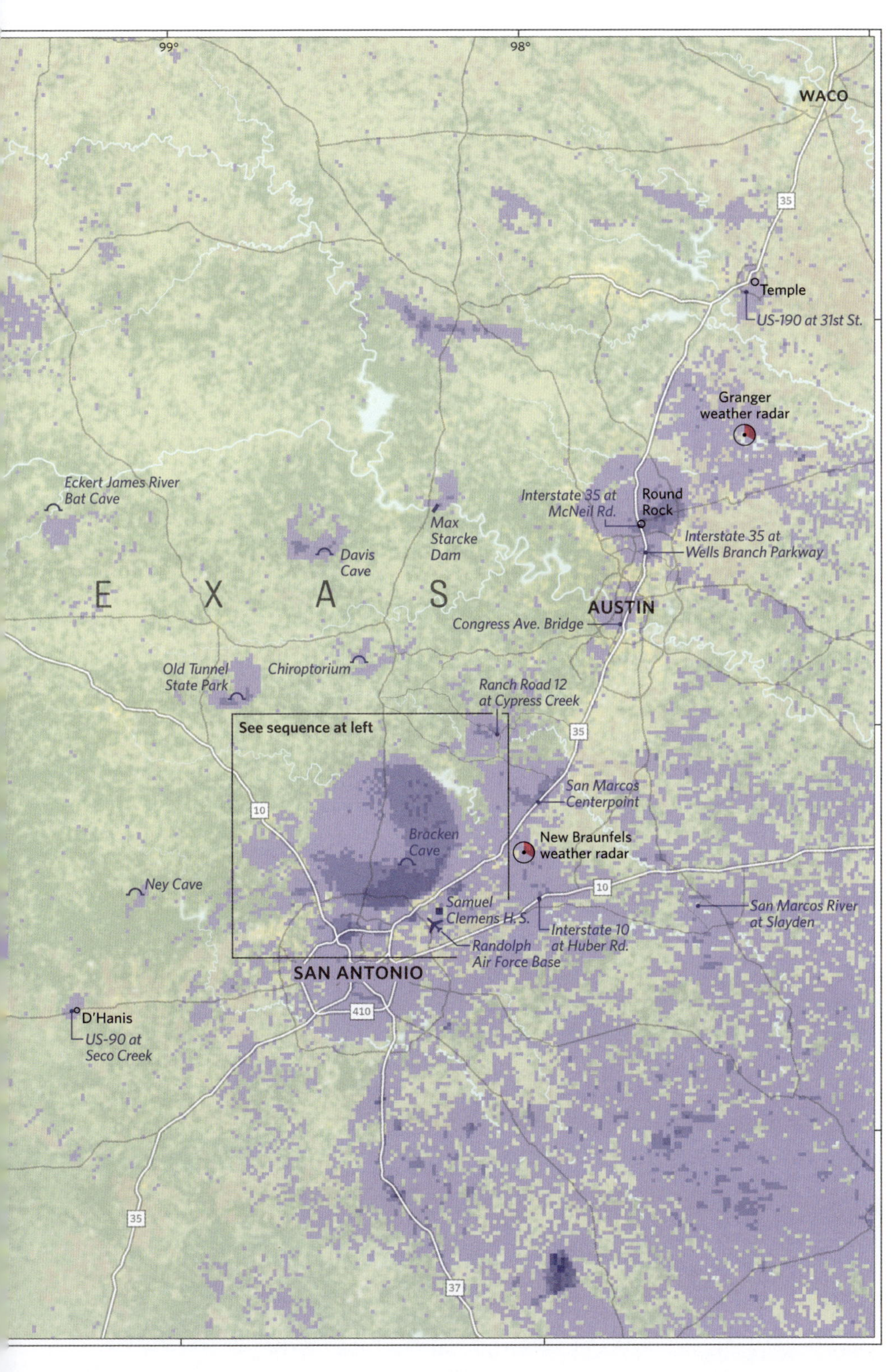
99°
98°
WACO
35
Temple
US-190 at 31st St.
Granger
weather radar
Eckert James River
Bat Cave
Interstate 35 at
McNeil Rd.
Round
Rock
Max
Starcke
Dam
Interstate 35 at
Wells Branch Parkway
Davis
Cave
T E X A S
AUSTIN
Congress Ave. Bridge
Old Tunnel
State Park
Chiroptorium
Ranch Road 12
at Cypress Creek
See sequence at left
35
San Marcos
Centerpoint
10
Bracken
Cave
New Braunfels
weather radar
Ney Cave
10
Samuel
Clemens H.S.
San Marcos River
at Slayden
Interstate 10
at Huber Rd.
Randolph
Air Force Base
SAN ANTONIO
410
D'Hanis
US-90 at
Seco Creek
35
37

3

E Pluribus Unum

Nations

IN THE BLEARY, predawn hours, the wan glow of hillside streetlamps above Caniçal harbor could still be mistaken for the edge of the world.

As I stood on the tower balcony of the *Funchalense* 5, a mighty cargo vessel, the crew was making final preparations to push off from the volcanic island of Madeira, destined for the open ocean. This North Atlantic archipelago, more than 500 miles southwest of the Portuguese mainland, once served as an outpost visited by early European navigators and Moorish nomads who floated over from Africa on dhows filled with silks and spices. Those who foolishly sailed beyond the island's western shore were presumed to fall off Earth's rim and into the abyss.

That all changed when explorers began regularly crossing the Atlantic, returning with stories of vast, unclaimed abundance. To sit in Madeira's bustling port during that time would have afforded a front-row seat to humanity's accelerated pillaging of the natural world. No longer relegated to the global periphery, Madeira was reborn as a strategic transit point through which the world's commercial goods flowed. Sugar, wine, wheat, and the island's namesake resource: *madeira* means "wood."

As global maritime trade accelerated, a shipbuilding boom brought the lumber industry to Madeira's shores, where it embarked on decades of mass extraction of highly durable trees from the island's lush forests, decimating the landscape. By the 1800s, this shortsightedness became apparent when Madeira's subtropical climate started attracting European aristocrats and expats, leading to an unexpected domestic building boom. In an ironic twist of fate, residents of what the Medici Atlas once called Isola della Lolegname—Island of Woods—soon resorted to importing lumber from abroad to build their own houses and burn as firewood in sugar mills and steamboats.

Eucalyptus and acacia trees started arriving from Australia and New Zealand. Jacarandas, Brazilian pines, and Tipuana trees from South America. Hardwoods were haphazardly swiped from all over the globe in a desperate attempt to replenish the island's depleted supply, with little regard for how the foreign trees might fit into the local landscape. Within a matter of decades, the eucalyptuses, having been abruptly transplanted to a place with no hungry koalas to keep them in check, started growing out of control, and the acacias smothered native plant life.

But an even bigger surprise soon arrived, likely on a trade ship from Argentina. Unbeknownst to the ship's captain, the trees they hastily snatched may not have been the only "invasive" species aboard.

WHEN HENRY BESTON first reimagined the animal kingdom as one composed of "other nations," he employed that metaphor in service of making a more fundamental point: that humankind is not the standard by which other species should be judged. That the sophisticated animal societies he glimpsed through his seaside cabin windows do not exist *for* us, or *below* us, but *alongside* us.

That worldview would have come as a surprise to some of the leading thinkers from the Age of Discovery. Even as Columbus was sailing the ocean blue, many of Europe's top scientists and Renaissance writers were still extolling the virtues of an ancient ranking system known as the Great Chain of Being. On the top rung of that hierarchy was, of course, God. Below Him were angels, followed by all of humankind, from kings and queens to commoners and thieves. Who could possibly be lower than the lowly thieves? This, the Great Chain said, was the domain inhabited by nonhuman animals.

At the time, engaging with such creatures was believed to hold implicit danger. The first world map that featured the word *America* on it, illustrated in 1507 by German cartographer Martin Waldseemüller, warned sailors about ferocious and unpredictable sea monsters. Human beings were out of their element in the open ocean, his map suggested, vulnerable among amoral hordes.

But as European explorers became increasingly confident in their ability to command the high seas, the mood rapidly shifted. A mere decade after printing his original map, Waldseemüller released his updated *Carta marina* (marine map). This time, gone were the references to brutish aquatic challengers, replaced by a sketch of King Manuel of Portugal triumphantly riding a sea monster off the southern coast of Africa, a metal horse bit jammed in its mouth. The message was clear: Fear not the wild animals of the world. They are no match for us.

This supremacist mindset was later evident in the introduction of the term *civilization*, which was created to distinguish between what Europeans deemed refined and advanced societies from supposedly primitive or savage ones.

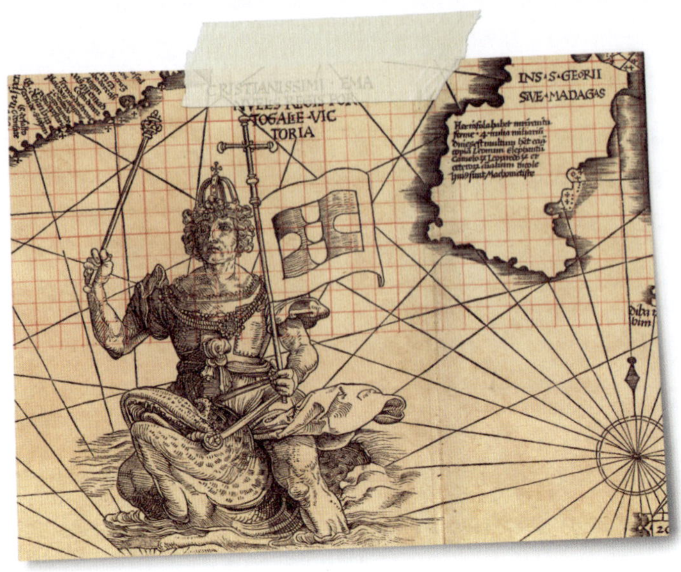

Suffice it to say that the criteria used to make such an assessment was plainly self-serving and based on a narrow set of European norms, so as to ensure that the user's own civilized identity was never in question. The Inca, for instance, earned enormous respect from the Spaniards for their great works of architecture, vast road system, and management of natural resources using advanced terracing and irrigation techniques, but were nonetheless deemed a "backward" society due to their lack of a written alphabet and non-Christian belief system. Among sedentary societies, nomads like Bedouins and Kazakhs, whose traditional oral storytelling was too often mistaken for myth, were viewed as living artifacts who were inherently less civilized. "Nomads have no history," a French philosopher later wrote. "They only have geography." Prejudices also prevented early European colonists from even acknowledging the impressive stone wall and tower constructions built in the southern African city of Great Zimbabwe, which they falsely assumed must have been the work of a long-lost white civilization. Other criteria, like equity and sustainability, were scarcely considered at all.

Over time, we have rightly expanded the definition of civilizations to include a wider variety of societies and practices, with an emphasis on complexity, resilience, and cultural richness rather than on superiority. But only recently have we started to even consider the notion that other animals may be capable of maintaining complex and resilient societies, acquiring and exchanging knowledge, and realizing achievements that go beyond our own metrics and abilities.

This human supremacy over the rest of the animal kingdom is one that Beston implicitly rejected by framing other species not as underlings tragically hindered by their own incompleteness, but as parallel societies with their own sophisticated life-worlds. In choosing the word *nations* to make his case, he also became the latest in a long line of writers to affix new meaning to a potent term.

Unlike the rigid confines of *countries*, nations are bound not simply by politics or geography but by common descent, traditions, culture, or language—links that connect disparate individuals who may never meet. Importantly, nations are also not fixed in time. They are amorphous, continuously shifting, merging, splintering, spawning new subcommunities, or sometimes disappearing entirely.

Within our own species, this fragmentation manifests in many forms. It's what empowers the Basque and Catalan people to distinguish their communities from those of their Spanish neighbors. The Navajo Nation spans parts of Arizona, New Mexico, and Utah but remains fundamentally its own thing—culturally distinct from the American states that now overlap it, and from the neighboring Indigenous communities with whom they share land. As recently as the 1940s, fewer than thirty people outside the Navajo Nation understood their language—a reality that enabled the famed Navajo "code talkers" to disseminate highly classified material on behalf of the Allies in World War II, confident that no eavesdroppers had the slightest idea what they were saying.

The Rohingya in Myanmar, Hmong of China and Southeast Asia, Kurds of the Middle East, and the Cornish nation back in Penryn are just a few other examples of populations that are linked through a shared history and broad cultural diaspora into which new people are born every day. Some have physical territory that they lay claim to. Others exist only in the mind.

It is in this spirit that Beston's vision of nonhuman nations takes on new resonance for me. His is a world that allows for a seemingly infinite number of distinctive animal societies, subgroups, and populations, "gifted with the extension of the senses we have lost or never attained, living by voices we shall never hear."

STARING OUT OVER the deck of the *Funchalense* 5, I realized how easily a stowaway could get overlooked. This freighter was longer than a football field, with only a handful of hardy crew members responsible for hundreds of identical shipping containers filled with bananas, oil, wood, and other miscellaneous goods destined for mainland Europe. The ship's manifest list had space for a couple dozen disembarking passengers, but on this week's voyage, I was the only one.

Or was I? Some travelers are so small, so unnoticeable, that they could easily slip under the radar.

Such was the case around 1858, when a trade ship pulled into Madeira from South America, whose crew was seemingly unaware that somewhere on board were an untold number of tiny Argentine ants.

Inadvertently plucked from their native forest floor, the ants soon found themselves transported to a foreign place unlike any they had ever known—5,000 nautical miles from their Southern Hemisphere home. Their arrival in Madeira helped spawn a transcontinental empire that now spans 3,600 miles from Portugal to Italy, populated by billions of individuals, creating what researchers have called the "largest cooperative unit ever recorded."

Rolling side to side in my sea cabin's walled bed during our three-day journey to the mainland, I passed the time between meals brushing up on the peculiar traits of the unassuming insects I'd come to visit.

Argentine ants (*Linepithema humile*) are distinctive in many ways, most prominently because they have demonstrated an unusually strong ability to instantly recognize any unannounced visitor as either part of their nation or not by assessing the chemicals on the guest's exoskeleton. Those visitors whom they identify as being fellow citizens of their community ("One of us!" "One of us!") are promptly welcomed and integrated into society. Those perceived to be foreign intruders are rapidly vanquished. It's an epic world-war clash, taking place less than a half inch off the ground.

This sort of with-us-or-against-us identification has caught many would-be competitors, like fire ants and spiders, off guard. In a head-to-head battle, such opponents would easily crush the tiny Argentine ant. But the Argentines' strategic advantage is not their size—it's their united front. By swarming en masse, they can overwhelm larger opponents and steal the territories of anyone who stands in their way.

This ruthlessness has earned the ants a vicious reputation among ecologists, who have variously branded them a "scourge in the Mediterranean," "unique among injurious insects," and a "household pest of the first rank." They have caused public hysteria at various points, leading at least one newspaper to claim, falsely, that the ants have "been known to eat a baby." A hostile nation if ever there was one.

But curiously, the ants had no such reputation in their ancestral homeland of South America, perhaps because they were kept in check there by highly aggressive native species and other natural enemies. It was only when they were erroneously swept away and transplanted to the North Atlantic that their footprint began to grow exponentially. Scientists have theorized that this is because once the ants found themselves in a foreign place, it became strategically advantageous for those of similar genetic history to link

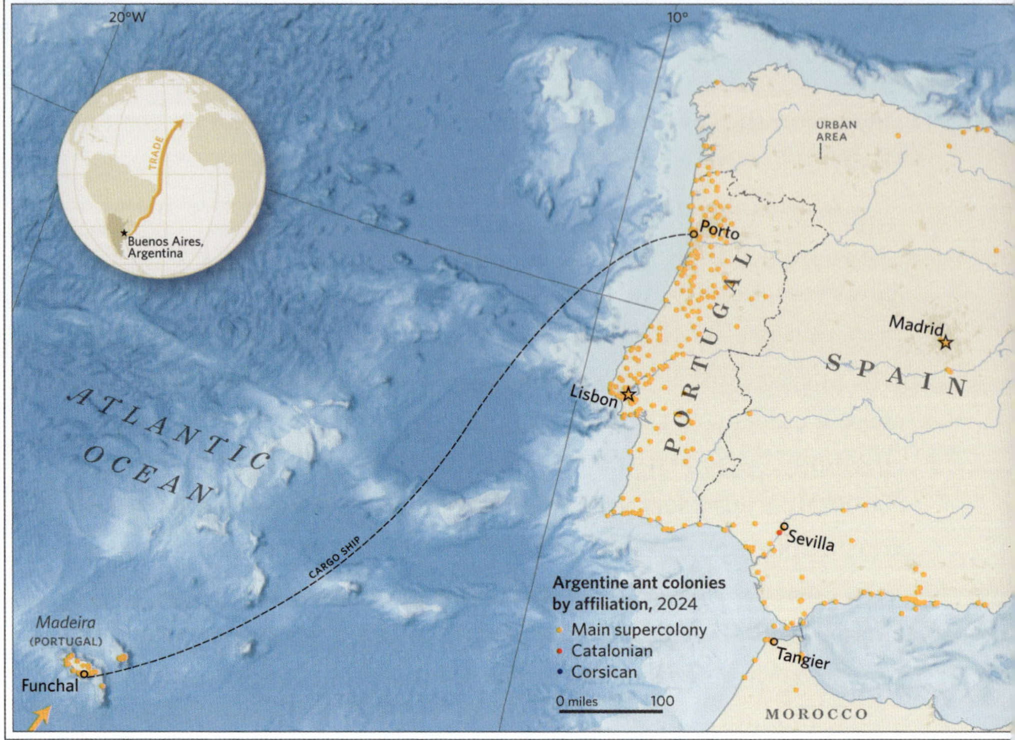

arms, so to speak, to collectively combat the unsuspecting native species.*

As I pondered this odd history and our heaving vessel approached landfall on the western shore of continental Europe, I grew increasingly conflicted about the narrative of the ants I was about to meet. If their arrival in Europe can indeed be traced back to a forced physical relocation by the extractive lumber trade or other human commerce,† theirs appeared to my eyes to be less a story of malicious invasion than of displacement and opportunism. A tale of animals who unexpectedly found themselves in a "New World" populated by ill-equipped competitors. Transatlantic colonization by mistake.

---

* While some similar behavior has been observed in the ants' native range, it's on a much smaller scale than in Europe, and fighting between colonies remains the norm there.

† Alternative theories about what the ships carrying Argentine ants were transporting include wine, sugarcane, or even South American tourists.

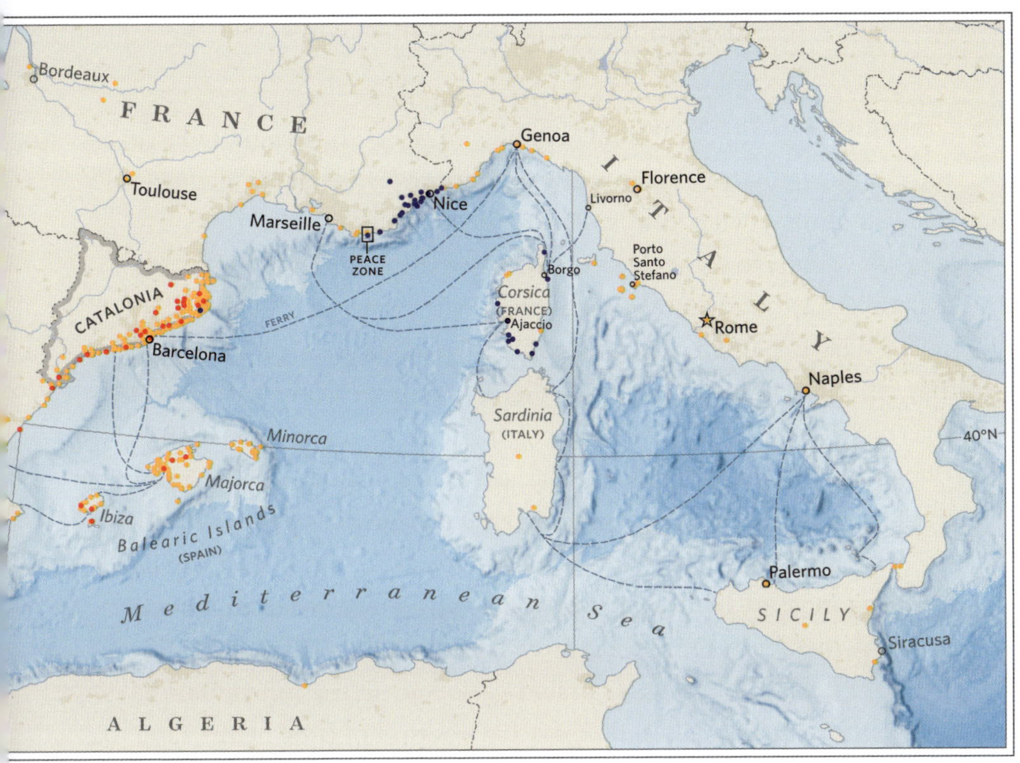

RECOGNITION OF ONE'S own kind is a fundamental attribute in the survival of nearly every species, including humans. After all, if animals were unable to intuit who is like them—who is a dolphin and who's not a dolphin, for example—the mere act of reproduction would be virtually impossible.

Species-level identification allows social animals to interact through a baseline set of shared norms, such as mutual recognition of a homeland, landmark, or territory, or common acceptance of boundaries and sovereign areas. ("This is my beaver dam. Yours is over there. Stay in your lane.") Among some species, shared communication systems—verbal or otherwise—can be used to convey useful information about food sources or perceived threats. When encountering an unfamiliar individual, such methods can also be used to announce, "This is who I am and where I'm from."

While Argentine ant supercolonies may be an unusually large and complex alliance, other animals also use a wide array of cues—visual, auditory, behavioral, chemical, or something else entirely—to identify who is, to some degree, similar or different from them, and "whom it may be advantageous to align with," says Max Planck ecologist Chase L. Núñez. Groupings and subgroupings, large and small, that differentiate beyond a species level and link certain individuals together. Determining whom it is beneficial to collaborate with is "one of the building blocks of societies," Núñez says. Indeed, "cooperation, no matter how constant or rare, *is* society."

Some animal societies are rigid in structure, with individuals spending most of their time among the same tidy group of friends or family members. Others are more fluid, based on their particular needs. In what are known as fission-fusion societies, individuals or subgroups will routinely split apart for a period of time before later reconvening.

But animal societies also do not exist in isolation. For example, while the thousands of beaver families in and around Pakwaw Lake each operate as autonomous units, given the proximity of their dams and surrounding resources, shared genetics and communication techniques of the species, and occasional interactions between neighbors, that population is composed of loosely but meaningfully connected social groups. So, too, with the bats in Bracken Cave, who engage and overlap with adjacent bat communities as part of an interconnected population. It's in these subtle network connections and shared histories within a population (or a network of populations), which link together disparate social groups from within the same species, that animal nations start to take shape.

Sometimes the opportunities for social interaction, group diversification, or cultural exchange are quite limited. The Devils Hole pupfish, for instance, exists in exactly one water-filled fissure in the Nevada desert. Every member of this species lives their entire life within a geothermal spring pool. Since recordkeeping began, its population count has never exceeded 553, and at some points has been as few as 35 individuals. Confined to a single area and living like an uncontacted tribe with virtually no prospects of independent expansion, theirs is a very small nation indeed.

In most cases, there is plenty of room for nuance and complexity to emerge, informed by a shared descent, geographical area, culture, language, or other time-deepened behaviors. But distinguishing where exactly network connections start and end can be a confounding process, replete with gray areas.

In 2024, Ross Andersen, a staff reporter for *The Atlantic*, penned a fascinating piece titled "How First Contact with Whale Civilization Could Unfold." It tells the story of Project CETI, which has raised millions of dollars to use AI as a means of learning sperm whale languages so that we can perhaps someday establish communication with them. This is a complicated task for many reasons, Andersen explains, not least of which is that the whales' nested social structures are immense. "About 10 whales swim together full-time as a unit. They will sometimes meet up with others in groups of hundreds. All of the whales in these larger groups belong to clans that can contain as many as 10,000 animals, or perhaps more. . . . Sperm whales meet just a fraction of their fellow clan members during their lifetime, but with those they do meet, they use a clan-specific dialect of click sequences called codas," he writes. "Their codas could be orders of magnitude more ancient than Sanskrit."

Alliances and connections between networked animal groups can also shift for unexpected reasons. Before Hurricane Maria struck the Puerto Rican island of Cayo Santiago in 2017, the nearly two thousand rhesus macaques who live on the island were described by researchers as "despotic and nepotistic," aggressively intolerant of the groups and individuals they competed with. When the hurricane devastated the island, destroying two-thirds of all vegetation, most of the monkeys somehow survived. "We don't know where they went or how they managed that," University of Exeter behavioral ecologist Lauren Brent told NPR. Then, once the storm passed, researchers noticed that not only did the monkeys sit comfortably alongside former rivals in what Brent described as "little puddles of shade" from the few remaining trees, but overall aggression dropped significantly, and monkeys were seen hanging out together at other times of the day, even when they had the choice not to. The world as they knew it had changed, so their society changed with it.

AFTER DISEMBARKING ON the Portuguese mainland, I bolted down the coast and crossed into Spain to meet up with entomology power couple Drs. Elena Angulo and Xim Cerdá. Having worked in the field for fifteen and twenty-five years, respectively, they know this region's insect landscape better than anyone.

As we drove along barely discernible dirt roads in Doñana National Park just outside Sevilla, we approached an area where cool sea breezes blend with the sandy scrublands—the ants' sweet spot, Angulo said. Unlike, say, caterpillars, who consume water by eating plant leaves, Argentine ants need fresh water to drink and carry back to their nests. Coastal moisture and humidity are critical to their survival, but an area also has to be dry enough for a land-bound insect to thrive without drowning. In other words, they crave that divine combination we call a "Mediterranean climate."

To demonstrate this point, the couple took me to a specific cork tree that they said is a reliable ant-nesting spot. Grayish, knotty, and crenulated with deep grooves in its bark, the tree was unremarkable and anonymous in this landscape. Even when I looked at it from

10 feet away, it appeared completely static, with only the whispers of sea breezes breaking the silence.

Inching closer, I started to notice what looked like shifting sand particles caught in the tree's bark—swirling rapidly as if trapped in a kaleidoscope. It was mesmerizing but indistinguishable. My eyes saw movement, but couldn't discern why.

Only when I was less than a foot from the surface did my eyes focus enough to see individual ants busily traversing the trunk in what looked like a choreographed dance routine. "My god, they're *tiny*," I said.

No more than three millimeters in length—about the size of a sesame seed—the ants are perceptible to us only when gathered in large numbers, which they almost always are. Lifting up a brick slab in the dirt next to the tree, Angulo revealed an entire ecosystem of what appeared to be thousands of shiny ants whirling around.

The ants will not travel far from this spot, she said. Maybe 60 to 100 feet, depending on their food resources. But they might interact with a neighboring nest, which then interacts with another nest farther away, and so on and so on, creating an informal network. In Europe alone, the ants are presumed to maintain millions of nests.

In the dry season, the ants will sometimes go farther afield to temporarily colonize a different area ("their beach house," in Angulo's parlance), before contracting back and huddling in their home base for the winter. Only if the beach house turns out to be a significant upgrade from where they have been living will they choose to stay long-term.

When the ants venture into new parts of the park and bump into another Argentine nest, they will be immediately welcomed in like long-lost sisters and brothers. In this corner of Spain, all the Argentine ants are part of what's simply known as the "main supercolony," meaning everyone from their species is a friend. It's this strategic

nonaggression that has enabled the ants to dominate ever greater swaths of territory, because they fight as one, rather than wasting any energy dueling each other.*

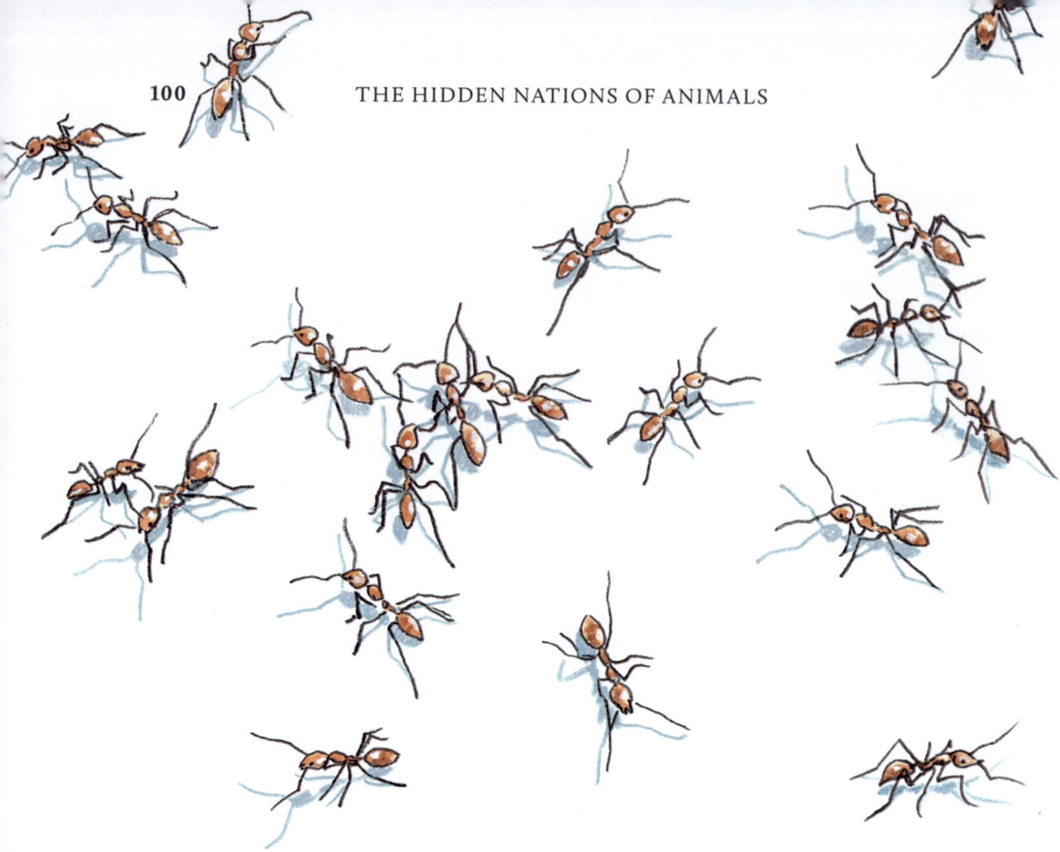

But if an Argentine ant scout happens to venture into a desirable location or food source populated by a different species, a chemical warning flare will be sent up—an alarm pheromone that is silent to our ears but instantly detectable to the ants' peers through powerful antennae receptors. Upon sensing the alarm, Argentine troops will arrive from far and wide to charge forward and invade the neighboring nest *Braveheart*-style.

In a 1913 USDA report simply titled *The Argentine Ant*, researchers wrote that if "a large food supply is discovered at a considerable dis-

* The exception to this internal harmony, Spanish researchers told me, is the Argentine ants' time-honored tradition of waging mass executions of their least reproductively prolific queens every May—an annual coup d'etat that enables them to optimize population growth within each community.

tance from the colony, a heavy trail of workers will soon be formed between the food and the nest, composed of many thousands of tiny insects, each busy carrying a load of the coveted material back to the nest or going out for another load. Sometimes they will construct a new nest in the neighborhood of the food supply. . . . In the course of a day or so this new colony will be thoroughly established, with a full supply of queens, workers, and immature stages, and will then be capable of supporting itself and increasing in numbers without assistance from the parent nest."

Near one of Doñana's field research stations, my guides drew my attention to what appeared to be the remnants of a recent clash. On the brick patio, we saw a smattering of gypsy ants* stumbling around stunned and disoriented. Scattered among the gypsies were small battalions of Argentine ants. Cerdá found this mixed-species scenario highly unusual. "Something just happened here," he said, squinting like a blood-spatter analyst. "My guess is that shortly before we arrived, the Argentine ants invaded a gypsy nest and displaced them, which is why the gypsies are wandering around randomly. They're homeless now."

Sauntering over to a nearby tree, Cerdá pointed out the nests of several different native ants who have so far managed to escape or repel the Argentines' wrath.

The first was a kind of tree ant whose chemical pheromones leave clearly demarcated rust-red pathways in the bark known locally as *autopistas de hormigas* (highways of ants). These routes are so heavily trafficked that between the ants' pheromones and their millions of microscopic footprints, it slowly eats away at the bark, carving permanent grooves in the tree and creating a network of tiny slot canyons.

* This common title has been the subject of some controversy, with the Entomological Society of America voting in 2021 to rename it out of concern that *gypsy* is derogatory to Romani people. European researchers don't appear to share this concern.

A similar situation, Angulo noted, was playing out in the sand nearby, where she pointed to a clearly defined trail blazed by a native granivorous (grain-eating) ant, no doubt leading to a reliable food source. Most ants prefer to avoid the scorching midday sun, she said, leaving their golden highway visible but vacant. Come sundown, this thoroughfare would be jam-packed with commuters.

As with human-built highways, heavily trafficked routes are subject to deterioration. Rain can not only destroy infrastructure, but also evaporate the chemicals the ants rely on to navigate their paths, effectively causing a washout. As a result, Angulo said, some of the granivorous ants are specifically responsible for keeping the highway clean, performing road maintenance for the benefit of the rest of society. If a highway eventually outlives its usefulness, either because the goods at the end of the road have been depleted or a more efficient bypass has been discovered, the routes will slowly disappear, reclaimed by the sands of time.

Argentine ants also have complex pheromone transit networks, Angulo said, but given how small the individuals are (even by ant standards), their trails are not generally visible to our eyes without special tools. To the ants themselves, they're clear as day.

So dependent are ants on these chemical highways that if researchers spray pheromones on the ground, the ants will follow the new trail precisely, like Homer Simpson chasing the scent of donuts. One researcher even managed to use a pen filled with concentrated pheromones to write the word "ANTS" *in ants.*

One thing ants and humans have in common is that we both loathe inefficiencies in our transit systems. In Europe and other places, Angulo noted, urban developers will often look to ant societies for inspiration on how we can optimize our own infrastructure and prevent traffic jams. Researchers have shown, for example, how during periods of heavy foot traffic, leaf-cutting ants will each carry smaller leaf loads, saving the medium and extra-large leaf fragments for times with lower congestion. They do this because having even one ant carry an extra-large load slows down traffic for everyone trapped behind them, creating an intolerable delay. City planners have pointed to such strategies in making the case for restricting highway access for large trucks and vehicles during peak periods.

Argentine ants also "show a remarkable capacity to deal with the many problems associated with building an efficient transportation system in a dynamic environment," researchers have noted. They will track changes in foot traffic and gradually phase out pheromone trails that have become redundant or irrelevant by redirecting routes toward the nests or destinations that are most in-demand. If there is a surge in traffic on a particular route, additional lanes will be formed to accommodate that need.

Such inspired urban planning lingered in my mind when I boarded the high-speed train to Barcelona for a scheduled meeting with researcher royalty. As we pulled out of the station at the peak of rush hour, Sevilla's cityscape dissolved into a blurred frenzy outside my window, and highway on-ramps overflowed across the land.

IF YOU DIG deeply enough, much of the modern-day knowledge of ants in Europe can be traced back, in one way or another, to Professor Xavier Espadaler. He spent forty-five years studying ant species and the complex dynamics of their societies before retiring in 2015. Throughout his career, he trained generations of researchers who have since carried on his legacy. Without exception, his former pupils speak of him with a kind of reverence, often deploying a nickname Espadaler wholly rejects: "the King of Ants."

"I am not so important. Nobody is important," he said wistfully.

"Everybody is limited in time. When you see millions of years of evolution, my short life of seventy-two years is nothing at all. I am an equal," he noted, raising a finger and leaning forward. "Just one."

Espadaler was credited with many breakthroughs over his decades in the field, including the first identifications of several ant species never before observed in Spain, which inspired others to follow his lead. But his work took an unexpected turn in 2002, upon the discovery of the main Argentine ant supercolony's biggest European rival: an insurgent group known simply as the "Catalonian supercolony."

With local logistical support from Espadaler, a team of Danish, French, and Swiss scientists was the first to conclusively show that what had at first appeared to be one giant supercolony stretching across Europe was, in fact, at least two groups, which demonstrated "extremely high" aggression toward each other.

The Catalonian supercolony is presumed to have a similar origin story, possibly arriving independently of the main colony's introduction on a ship transporting cork between Madeira and Spain. Perhaps as the result of a distinct introduction from Argentina into Europe, this second wave of arrivals did not bear the same chemical signatures as the main supercolony, meaning that they were viewed not as brothers and sisters, but as foreigners. The rise of a competing Argentine ant supercolony soon created a transcontinental collision. In 98 percent of cases, clashes between the two groups were fatal.

After the initial discovery of the Catalonian supercolony, Espadaler went on a tear. He drove up and down the Mediterranean Spanish coast conducting research to get a better sense of exactly where the supercolonies' borderlines were. The result was a comprehensive map that revealed widespread territorial disputes, street fights in

Barcelona, and rival ant nests that showed no genetic overlap despite sometimes being a mere hundred feet apart.

Clashes between Argentine ant supercolonies can be epic in scale, at least to the participants involved. University of California San Diego ecologist Dr. David Holway once documented what he described as "a World War I battlefield" along a fault line between two supercolonies. "You'd drive up and, before getting out of the car, you'd see the piles of dead ants," he told *The San Diego Union-Tribune*, describing the pitched battles along the borderline as "trench warfare, with a lot of mortality and not much movement."

The fact that this particular clash emerged in Catalonia, a corner of Spain renowned in human society for its vibrant independence movement, is merely a coincidence, Espadaler says. Nobody could argue otherwise "without quite a few beers." But curiously, when a French researcher named Dr. Olivier Blight reported in 2010 that he had discovered a third Argentine ant supercolony, on the Mediterranean island of Corsica—itself a hotbed of independence-minded fervor—Espadaler was among the researchers to dispute Blight's findings.

"I cannot accept it," he said, noting that while Blight showed that Corsican ants routinely attack both the Catalonians and the main supercolony they are presumed to have broken off from, at a genetic level, they appear to be virtually identical to the main one. By contrast, the Catalonian supercolony has forked off to such a degree at both a genetic and behavioral level that Espadaler believes they are different groups in ways that truly matter.

"I would fight for my son. I would fight for my daughter, for my family. For their genes," Espadaler said. "It also happens in wolves, sheep, fishes, in ants—everybody! But I will not fight as easily for the sons of my cousin, because there is only one-sixteenth of my genes there. It dilutes very rapidly. In my son, it is fully half. This is pure biology."

To his mind, if the Corsican ants' genes are identical to the main supercolony, they are one.

STEPPING OFF A train in Marseille, smack in the middle of the ants' European territory, I hopped into the first taxi I saw. "*Parles-tu français?*" the driver asked.

"*Non*," I said, pointing to a sheet of paper with my guesthouse address.

"*Italiano?*" he surmised.

"American," I replied, prompting a sarcastic guffaw that seemed to translate to "Ugh, useless."

After an awkward ride to my destination, I met up with Dr. Blight, discoverer of the disputed Corsican supercolony, and we embarked on a walking tour through the ants' preferred downtown haunts, almost all of which were city parks and ornate museums. For Blight and other ant researchers in Europe, iNaturalist and other crowdsourced apps and websites have become useful for verifying new supercolony sightings in urban settings and expanding our understanding of where the ants were last spotted or might be headed next. In some cases, he said, ant populations have disappeared entirely from parts of Marseille, for reasons he cannot confidently explain.

As with the Catalonian supercolony, Blight said, the origin story of the Corsicans remains unresolved. He believes there are two possibilities: an independent introduction event on the island or a drift that branched off from the main supercolony.

Intriguingly, Blight said that his work has also shown that as a collective unit, each of these three supercolonies has what he describes as its own "personality," informed by historical events. "The main supercolony is more aggressive, personality-wise," he noted. Their behavior profile is likely part of how they managed to dominate thousands of miles of Europe in the first place, but the supercolony's earlier victories have also made it more assertive in exploring new places than the Catalonian colony, or the Corsicans, who, he's noticed, generally stick to the areas they're in.

When I told Blight over lunch at a streetside café that I was headed to Corsica that evening on an overnight ferry, he marked up several locations on my map where he suggested I go to find Argentine ants. He would be curious, he added, to know if they are still in the same places they were during his last visit.

Blight is the first to admit that as non-ants, we can find it a challenge to make sense of what's really happening in their worlds. One of his colleagues spent four decades studying a single ant species, amassing a vast research bibliography that seemingly outlined every aspect of the ants' behavior. Only after the researcher's retirement was it discovered that, in fact, the ants he had been studying belonged to two different species. "Was he at least the person to discover the second species?" I asked, seeking a silver lining. He was not, Blight lamented, shaking his head.

This has become an increasingly common scenario. In 2024, *The New York Times* published an article titled "What Is a Species, Anyway?," which outlined the exasperating debate about where one kind of animal ends and another begins. The piece cites a 2021 survey, which found that "practicing biologists used 16 different approaches to categorizing species. Any two of the scientists picked at random were overwhelmingly likely to use different ones." The giraffe, the *Times* notes, was once viewed as a singular species, but was later divided into at least four distinct subgroups, each of which looks and behaves differently. Similar debates are underway to splinter orca whales into multiple species, based on distinctive behavioral, physical, and genetic differences, rather than lumping them all together. In these and many other cases, the animals themselves notice such differences long before we catch on.

In his own research, Blight has had to continually reevaluate his conclusions to try to make sense of what he's looking at. In 2013, he was part of a French research team that discovered what he described as a "peace zone" on the borderline between the main supercolony and the Corsicans. This zone features a site in southeastern France where aggression between the two groups was conspicuously absent, despite the mutually antagonistic colonies rubbing right up against each other.

Blight's leading theory for this is that, at some point, a small number of male workers were, perhaps inadvertently, welcomed into the opposing community—something possible only because tensions between the two groups are generally milder than with the Catalonians. Those males then had sex with the females from the other side, over time creating "colony fusion" by diluting the ants' natural aggression toward one another. In other words, through generations of reproduction, they started to see bits of themselves in their former enemies and became more tolerant as a result.

To Catalonia's King of Ants, who has cast doubt on the existence of the Corsican supercolony from the get-go, the notion of a peaceful buffer zone just serves as further proof that all these ants are, in fact, part of the main supercolony. But as we gazed out over Marseille's majestic port from a high peak in Parc Émile Duclaux, Blight remained confident in his conclusions. Watching the lines of tiny tourists bustling along the waterfront, he hypothesized that many other "peace zones" likely exist all along the borderlines of Europe's ant supercolonies. It's just that nobody is looking.

AFTER SCURRYING OFF the Corsica Linea ferry armed with Blight's marked-up map, I zipped out of the parking lot in a rented Suzuki Swift, relishing each shift of the manual transmission.

As I approached Plage de Capitello, a coarse beach just south of the airport, I was greeted by an imposing steel and stone monument that declared: LE II JUIN 1793 C'EST D'ICI QUE NAPOLÉON BONAPARTE QUITTA LA CORSE EN BATEAU AVEC SA FAMILLE POURCHASSÉ PAR SES ENNEMIS. A gloating marker of the spot where Napoleon, an Ajaccio native later branded a traitor to the island's struggle for independence, was chased into the sea by Corsican nationalists.

Walking past rows of bronzed holidayers, I headed for a slender cut of sand that divided azure waves from emerald-green shrubbery, creating a tricolor cascade. Channeling my inner Olivier Blight, I determined that this spot looked like prime Argentine ant territory.

In the shadow of a sunbaked stone wall, I squinted to look in its cracks for signs of life. Glancing up at the late-morning sun, I triangulated where the coolest spots on the ground might be, turned to the greenery's outer fringe, and pulled back the brittle shrubs to get a clearer look at the dirt below. Empty. Nothing but what was, this time, just shifting sand.

As I walked along footpaths between the dunes, I spotted a few fat, redheaded insects wandering around almost as aimlessly as I was. Far too bulbous and disorganized to be Argentine.

A hundred feet farther inland, I came across a small hole in the ground where several dozen black ants were stockpiling foraged goods. They appeared to be connected to a longer line of ants who were slow-walking along a concrete curb. I sent Blight a few still photos via WhatsApp, along with thirty-eight seconds of raw footage of them milling about. "AAs?" I asked insecurely. "They are not AA," he curtly replied. "AA workers do not differ in their size. Looks like

Tapinoma." Chirping birds watching from a nearby tree offered what sounded like stifled laughter.

I drove farther down the coast and explored a neighboring beach, thinking that perhaps the ants had moved southward since Blight's last visit. I wandered all along the seaside spots that seemed right up the ants' alley. Nada.

Discouraged, I got back in my diminutive coupe and rocketed two hours eastward on the paved ribbons that crisscross Corsica's vertiginous core. As I whipped past quaint medieval villages dotted with red-roofed chapels and balding peaks, the westerly winds seemed at times to be pushing me skyward. I slowed down only periodically to admire what are surely some of the Mediterranean's most picturesque vistas. Timeless panoramas of rugged cliffs and deep valleys, disrupted by the spray-painted scrawls of LIBERTA! on the roadside barricades.

Cresting over the central divide, I raced downhill toward the east coast's shallow lagoons, guided by Blight's scribbled notes, which identified a skinny spit of land east of Borgo that has been densely populated by Argentine ants for at least thirty years. If I can't find Argentines there, I thought, something is terribly wrong with them or me.

The air was noticeably stuffier on the eastern side of the island, sheltered by the mountains and unpunctuated by sea breezes. After pulling into a campground parking lot, I walked along labyrinthine pathways, staring downward at its dirty edges. Fifteen minutes of fruitless wandering later, I felt a sinking sense of despair.

What was I doing wrong? Or was Blight's research simply out of date and the Argentine ants had been eradicated or forcefully ejected from Corsica like so many others before them?

With my hand on my brow, I took in a 360-degree view, looked out into the brush, and spotted a dry, notched tree in the distance that looked vaguely reminiscent of the craggy ones I had seen back in Doñana. The spiky branches lashed my exposed legs as I pushed my way through the thick bushes, but I persisted. (If you're going to "take Vienna," take Vienna!) When I got within a couple feet of the tree's crusty exterior, I spotted a glimmer of motion. Then another one. I fixed my vision on the grooves between the bark, forcing my eyes to autofocus on the ambiguous movement. Only then did they suddenly appear like a Magic Eye illusion, frenetically scurrying up and down the thick trunk.

I immediately logged a video clip and sent it to Blight. "These have to be AAs, yes?" I wrote. I watched anxiously as the app's blue check marks signified that he'd received the message and viewed the footage. Then he started typing, and the confirmation came: "Yes! :-)"

"*La victoire!*" I responded, exhaling deeply.

Circumnavigating the tree with a renewed sense of confidence, I began following the ants' trails down to the dirt. Squatting lower revealed that they were indeed fanning out in all directions, traversing bustling thoroughfares that expertly navigated around stones and other natural obstacles.

Stung by a twinge in my knee, I abruptly stood upright to realign myself. At full stature, I glanced down, but again saw nothing but dirt. With my head cocked, I returned to a squat, and the ant highways reappeared, resembling a downtown artery viewed from a news copter.

How many animal societies surely must exist in this zone beyond our visual perception? I wondered. Massive in population density, but so small in human terms that they barely even register. The smallest ants are so tiny, so many times smaller than Argentines, that E. O. Wilson noted an entire colony of them "could dwell comfortably inside the braincase of the world's largest ant." Our eyes are woefully unprepared to appreciate such inconspicuous abundance.

I began to repeatedly stand and squat, grinning wildly in this patchy field as the ants' world ethereally appeared and disappeared from my view. To anyone nearby, I'm certain I looked like a gym rat who had gone completely mad.

Having tasted personal success, I strutted confidently back toward the main walkway, pausing every few feet to stop, drop, and look around for ant streams hiding in plain sight at lower elevations.

As I approached my car, I crossed paths with an elderly French couple out for a midday stroll. Still high on adrenaline, I felt a strong urge to blurt out, "There are Argentine ants over there!"—a remark that almost certainly would have been lost in translation.

FROM THE EAST coast Corsican town of Bastia, I hopped a whirlwind ferry followed by a forgettable train from Livorno to Florence, arriving in the middle of the night. By the next morning, I was in a car on my way into the central Apennine foothills with Italian researcher Alberto Masoni. My mind must not have kept pace with the rest of my body because even as our elevation climbed, my head felt deliriously stranded at sea.

With Tuscany's famed vineyards propagating below, we floated up into the thick tree cover of Parco Nazionale delle Foreste Casentinesi, high above the Argentines' territory. Pulling over on the roadside, Masoni pointed uphill, where we'd begin our short trek. About ten minutes later, we approached the first mound, maybe five feet tall.

I asked Masoni how many ants might live in this bulbous dome.

"Half a million," he nonchalantly estimated.

"And that one?"

"Same."

Surveying the hillside, dotted with mound after mound after mound, I did some quick back-of-the-envelope calculations, which revealed the total population in this area to be roughly . . . 100 million? Could that possibly be right?

"Somewhere around there, yeah," he said. A fresh wave of lightheadedness overtook me.

These giant haystack-looking mounds were created by red wood ants (*Formica paralugubris*), intentionally transplanted by researchers in the 1950s in an effort to push back on other local insects they viewed as pests. Many wood-ant nests were moved here from the Alps, of which three or four survived. Since then, they have expanded to at least six hundred mounds in this national park alone. Masoni has mapped them all.

Each mound contains hundreds of chambers, tunnels, and rooms, Masoni said, some of which are reserved for specific times of the year. In winter, for instance, the ants will cluster themselves in the lower chambers of the nest, which are connected to the outside by tunnels that lead to the top. The heat produced inside the nest structure is sufficient to keep the mound tops snow-free.

These forest cities are also interconnected by bustling pheromone highways, which facilitate free transit between central hubs. As I inched closer to a mound's surface, a dizzying operation was revealed as a seemingly infinite number of individuals popped in and out of various exterior doorways, industriously transporting and exchanging raw materials to and fro.

Like the Argentines, these wood ants tend to prey on and push out other insects who stand in their way, though thankfully the two species inhabit different elevations, meaning they do not overlap, at least for now. "If they did," Masoni said, chuckling and waving his hands in a *fuhgeddaboudit* tone, "by god, could you imagine?"

THE HILLS AROUND Florence serve as a kind of natural terminus for the Argentine ant. An insurmountable physical barrier that prevents further intrusions into the Apennines and beyond.

As in the other places I visited, the Argentine concentration in central Italy is primarily clustered along the coast, which is why I chose to end my time in Europe in a wetland area called Laguna di Orbetello di Ponente. When Masoni and I pulled into a rugged campground with his mentor, Dr. Giacomo Santini, it took us no time at all to locate dense streams of Argentines rushing around. "Traffic jam!" Santini chirped. As in other places of high concentration, he said, all the pheromone trails in this campground were connected as part of a larger system.

Pointing at a line of Argentines flowing out of a hole bigger and deeper than they would ever dig for themselves, Masoni said that this appeared to be just the latest example of the ants supplanting a larger native species and expropriating their home.

He attributes such victories to what are known as Lanchester's laws. Devised during World War I, these mathematical models are used by military strategists to game out the odds of victory when one army clashes with another. In a traditional "linear law" situation, where combatants engage in one-on-one battles without the ability to gang up on opponents, the sheer numbers of individuals involved on either side often determine who will ultimately win a war of attrition. But in an all-against-all scenario where groups can asymmetrically concentrate their firepower on certain targets, the "square law" rule kicks in and a group's strength increases to their population count squared. This gives a significant advantage to armies that coordinate their attacks collectively as a single unit. That dynamic, Masoni said, appears to benefit the Argentine ants everywhere they go.

It takes only about ten workers and one queen to start a new Argentine ant colony—a micro-community that could fit on a human fingernail. Shortly before my arrival in Europe, Santini said, Argentine ants had been found on Réunion Island—the first time they'd been identified anywhere in the Indian Ocean, no doubt assisted by human boats and planes.

In the 150 years after Argentine ants first appeared on Madeiran shores, they started showing up in Algeria, South Africa, Morocco, Bermuda, Hawaii, the American South, Indonesia, and the Philippines, among other places. By the 1990s, they had reached Japan, the United Arab Emirates, and New Zealand. In 2005, they were even found in North Korea. Today, no continent but Antarctica is devoid of their presence.

In some pockets, like Southern California, Argentine ants have systematically wiped out native species across the region. When I visited a canyon preserve north of San Diego with David Holway, the scientist who previously documented the ants' "trench warfare," he said that he used to come to that specific preserve for the expressed purpose of observing native ant species because it was such a hotspot of diversity. Now it's only Argentines as far as the eye can see.

The majority of these expansions have come from the same main supercolony that is also dominant in Europe. Remarkably, those ants have managed to retain their supercolony recognition and shared chemical language despite their ever-growing global diaspora. When ants from Europe's main supercolony have been introduced to more recent arrivals in Japan, they still greet each other like old friends—members of what Espadaler and other researchers have dubbed the Intercontinental Union of Argentine Ants.

THERE ARE SOME isolated spots where native ants have appeared to hold their own. On Corsica, Blight told me that a local species, *Tapinoma nigerrimum*, had managed to repel further intrusions into the island by leveraging its home-field advantage over the Argentines' comparatively smaller numbers. That is, until he later discovered that what he thought was a single native species was actually intermixed with three "invasive" species, making it unclear to him who's fighting who.

On a global level, the Argentine ants' crusade continues largely unimpeded. Blight says that few things are capable of slowing or stopping their expansion, and humans are not among them—indeed, we are much more likely to aid their expansion than stamp it out. Back when I was in Barcelona, Espadaler remarked to me that while Argentine ants don't need human beings to thrive, we often unwittingly aid their introductions into new places. Densely populated urban and suburban areas, he noted, are not natural settings for ants to live in, especially those traditionally accustomed to the riverine wilds of South America.

But during his decades in the field, he observed that Europe's Argentine ants appear to have learned that when humans develop an area, one of the first things we do is clear-cut the native grasses and shrubs and spray them with pesticides that kill any and all insects who were there before us. This slash-and-burn approach creates a clean slate upon which humans can then grow bright green lawns and install manicured landscaping, complete with fountains and sprinklers.*

For the opportunistic Argentines, this is great news. By installing year-round water features, our downtown areas manage to perfectly simulate the coastal conditions the ants love. Plus, by carpet-bombing the native insects, we inadvertently open up a power vacuum in the insect world that the Argentines can swiftly fill, allowing them to seize new territory without having to fight a single battle.

"You tell me," Espadaler asked, "who is the invader in that scenario?"

---

* Interestingly, some anthropologists now argue that beavers performed a similar service for early humans by reengineering and taming landscapes overrun by glacial melt in a way that enabled hunter-gatherers to better navigate large areas.

Still, there remains a distinct possibility that the Argentines will encounter a new type of opponent they have not yet faced off against. Some estimates have pegged the total number of ants on Earth at 20 quadrillion (that's 20,000 trillion), so for all the Argentines' intercontinental colonization, they remain a relatively small drop in a very big bucket.

Many species boast unique attributes that give them strategic advantages over their competitors. One study found that in the heat of battle, Florida carpenter ants will act as field medics, performing leg amputations on their injured nestmates and licking the wounds clean to remove bacteria, which enables at least 90 percent of amputees to return to duty.

Only a few weeks before I arrived in Italy, a red fire ant named *Solenopsis invicta*—also originally from South America—was spotted in Europe for the first time, landing in a suburban part of Sicily. The entrance point, local researchers said, was almost certainly the city's commercial port.

Blight believes that such discoveries are just the tip of the iceberg. There is often a lag of several years between when a species arrives in a new area and when we notice it for the first time. Odds are, he said, this fire ant has actually been in Europe for some time, living just below our radar.

Not all territorial shifts are accidental either. Increasingly, they are born of necessity. In many parts of southern Europe, climate change is wreaking havoc on the ants' preferred landscapes through longer periods of heat waves and drought. Three-quarters of Earth's land has gotten drier in recent decades, according to the United Nations, including more than 95 percent of Europe. Not once during my weeks-long journey from Madeira to Florence did I get rained on.

Reflecting on this fact, I was reminded of a story I had read in the newspaper by two climate reporters who had traveled through dozens of villages across Mesopotamia, a corner of present-day Iraq hailed by some as the "cradle of civilization." In recent years, the journalists wrote, accelerating desertification and a squandering of finite resources were prompting many human communities to relocate by literally "dismantling their homes, brick by brick, piling them into pickup trucks—window frames, doors and all—and driving away."

Such will be the challenges faced by many animal nations as desperate migrant groups increasingly move to a shrinking number of suitable habitats. This grand ecological reshuffling may, in turn, increase the risks for international clashes and new threats like accelerated disease spread.

Ironically, for all their collective strength, Argentine ants are acutely susceptible to such threats because their gene pool is so narrow. "There are some places where there's more diversity in a single tree than in an entire supercolony," says University of California, Berkeley, ant expert Dr. Neil Tsutsui.

That's a problem for them, because in other species with a wide variety of genetic diversity, new stressors impact some but not all individuals, at least a portion of whom would have already adapted to life in a different climate. But Argentine ant supercolonies are so uniform that changes affect the entire society equally, potentially leading to vast population implosions. In France, Olivier Blight believes this could help explain why Argentine ants have disappeared from some of their familiar haunts in Marseille. The bigger an Argentine ant nation gets, the harder it falls.

The ants' next chapter is being written right now, Tsutsui says, and they may not like how it ends. Only a couple of decades after arriving in New Zealand, where the ants' genetic diversity was among the lowest of any place they have established themselves, researchers noticed that the Argentines have suddenly begun to recede, collapsing by as much as 40 percent in many areas.

Word must have spread quickly, because as their numbers fell, displaced native ant communities swiftly returned, filling the void.

4

Below the Surface

Layers

"YOU KNOW WE have hippos here, right?"

"Yes, I'm aware of that," I brusquely replied, surveying the pockmarked soil amid a torrential downpour.

"Elephants too. Hyenas . . ."

I didn't bother to respond further, but politely nodded so as not to offend my guide or his machine-gun-toting scout, whose eyes remained fixed on the swaying trees in the distance.

It wasn't even supposed to be raining yet. There were still a couple of weeks left in the dry season, but you wouldn't have known it looking at the small peaks of moist earth popping up through the fallen leaves.

I could appreciate my escort's confusion. Unlike the stampede of tourists who arrive in this lush corner of Zambia's Kasanka National Park to rubberneck charismatic megafauna, I appeared—to everyone's bemusement—singularly focused on an animal whose presence is not seen but inferred.

Gawping at a maze of wet mounds before me, subtle signposts in the dirt, I surmised that for the giant mole-rats burrowing beneath our feet, today's thunderstorm was an opportunity.

STRICTLY SPEAKING, MOLE-RATS are neither moles nor rats. They occupy their own category of rodent, found only on (or more precisely, *in*) African soil. Until very recently, little was known about the scale of their wholly subterranean worlds except that they're down there. Somewhere. In the dark.

This, despite the fact that mole-rats have been about a foot below humanity's feet since the dawn of our species, to say nothing of the millions of years before that.

The giant mole-rats found in northern Zambia and in parts of neighboring Angola and the Democratic Republic of the Congo are about the size of a puppy. Smaller ones are closer to a squirrel. Kasanka reportedly has both kinds, and the fresh mounds dotting the landscape provide a glimpse into where they're going and where they've been. The wetter a pile is, the more recently it was excavated and pushed through the surface to make room for a new tunnel. But for millennia, it remained anybody's guess how exactly their underground networks were connected or structured.

Convinced that there was much more to know, in April 2009 a team of Czech researchers arrived in Zambia's Copperbelt region with the aim of producing the most comprehensive map to date of life in mole-rat society. Using a combination of radio-tracking technology and old-school excavation, the researchers unearthed a discovery staggering in scope, even to them.

Only after more than a month of digging could the scientists start to see the big picture.

Their mapping work, the largest of its kind that had ever been published, revealed a pair of mole-rat compounds that featured nearly two miles of expansive tunnels, dotted with family sleeping chambers and storage caches filled with wild plants harvested through the ceiling. Deep tunnels—informally known as "highways"—linked far corners of the complex and facilitated easy access to distant foraging areas. A completely self-sufficient bunker, insulated from the outside world.

Living inside the tunnels were at least fifteen individuals from two distinct families. Mole-rat family sizes in Zambia can vary widely by species, with some gathering in groups of more than twenty. Farther north in the Horn of Africa, the naked mole-rat, a smaller and more distantly related evolutionary cousin, can sometimes boast hundreds of individuals in a single burrow complex. Other mole-rats prefer to live reclusive, solitary lives, often initiated by branching off a single tunnel from a family complex before quietly building a thick soil wall behind them and starting life anew a few feet away.

Once built, mole-rat tunnels can last for years and possibly even decades in drier regions. Much of this construction takes place during the rainy season across central Africa's miombo woodlands, when soils are most shapable.

Deep below the earth's crust, the concept of day and night is largely irrelevant, so researchers believe mole-rats operate on whatever work and sleep schedule is needed to get their daily tasks done. Studies suggest they may have four or five bouts of activity per day, with each lasting a couple of hours. Given the variety of species across Africa, it's not uncommon to have overlapping tunnel networks at different depths stacked vertically, like an inverted condo tower.

As claustrophobic as this all may sound to us, mole-rats are optimized for life in their lightless tube world. With only a handful of places wide enough to fully turn around—something they do by folding themselves in half and pushing their head between their back legs—they have perfected the art of running backward, shut-

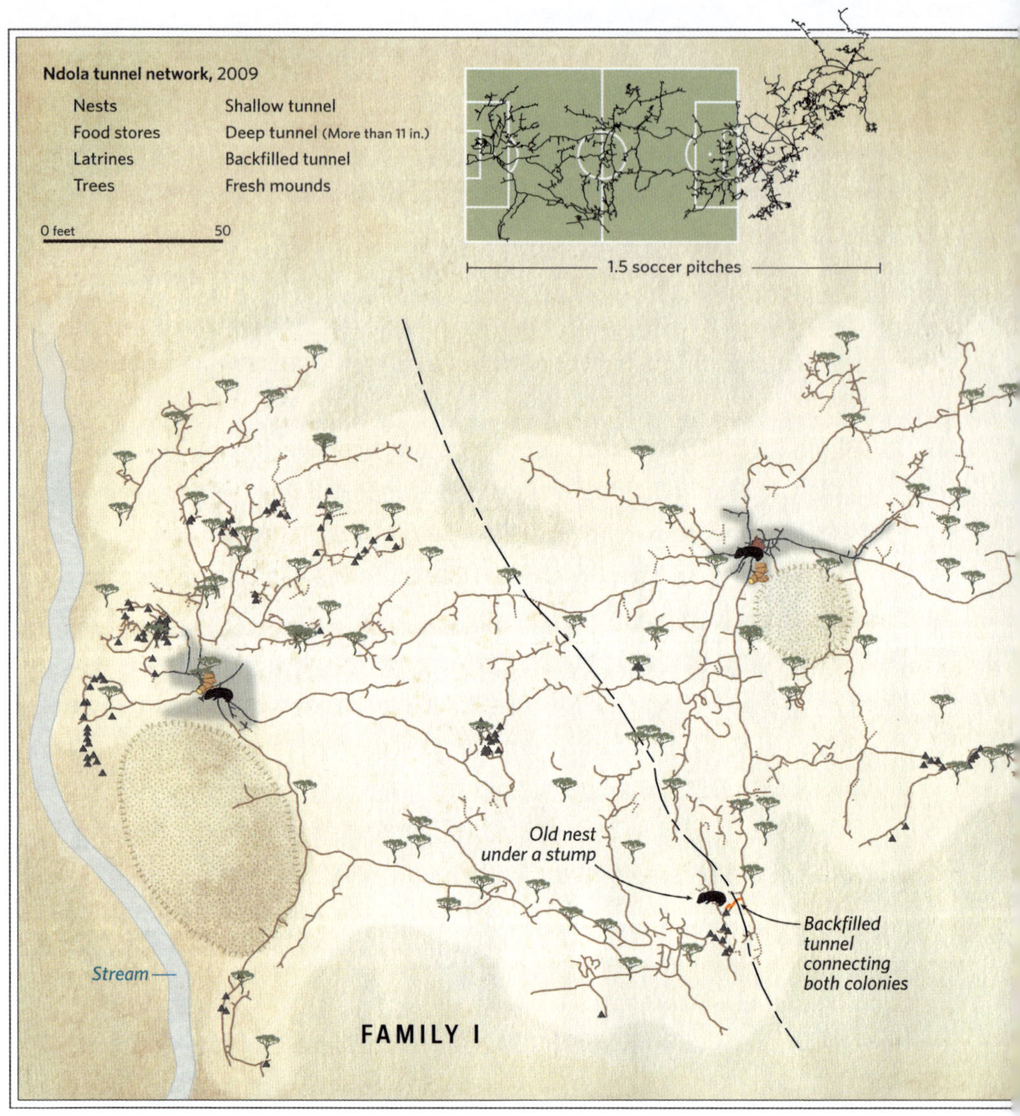

tling around like subway railcars. When females are pregnant, their spines expand, enabling them to make room for babies by growing longer, not wider.

The roots that poke through the ceiling of their tunnels provide all the sustenance they need to thrive. Plants with tubers and bulbs like wild ginger, peanuts, sweet potatoes, cassava, and a fleshy type of cucumber are among their favorites. Their four fang-like teeth

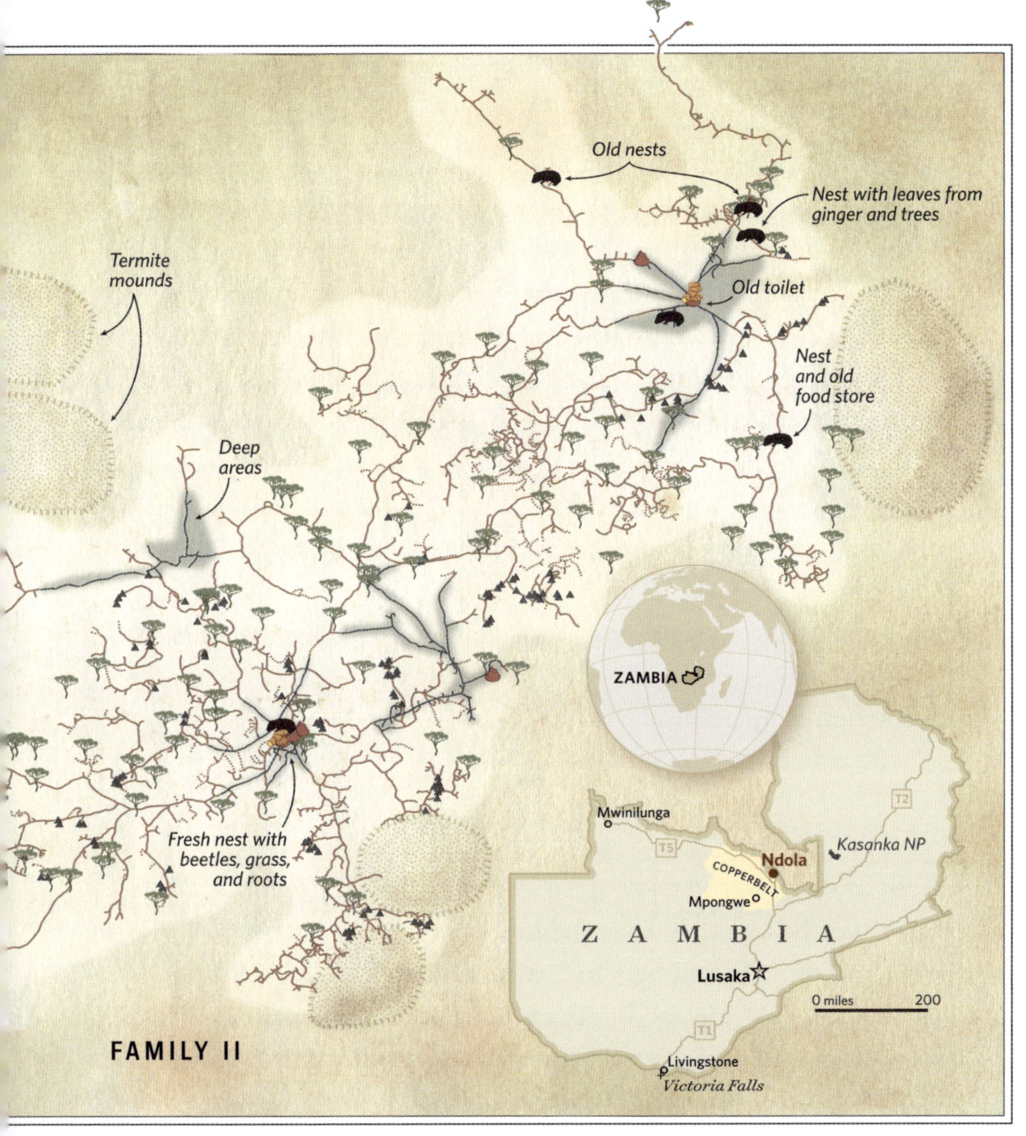

are, rather unsettlingly, not located inside their mouth, but instead protrude from the skin of their upper and lower lips, to prevent them from swallowing dirt while digging. They can also move their bottom teeth independently like chopsticks. Due to the lack of fresh water underground, they rely on the moisture of the plants they eat, which their bodies then metabolize into water.

Social mole-rats communicate with each other through vocalizations that carry a distinctive dialect, dictated by their queen. The queen's reign is a long one, with some mole-rats living for up to thirty years. If the queen happens to die unexpectedly, or there is what researchers call an "overthrow," in which a new queen tries to take charge—usually a daughter or an unrelated female who moves in—the entire community will change its dialect overnight to whatever the new queen says.

The only mole-rats who don't have a vocal repertoire are the solitary ones, because, well, there's nobody around to talk to. That's not to say that they don't communicate with the outside world. Solitary mole-rats are obsessive drummers, banging their feet up to twenty times per second to send shock-wave messages through the dirt in the hopes that perhaps, someday, a potential mate sealed away in a nearby tunnel might heed their call.

THROUGHOUT OUR MILLE-FEUILLE world, animals settle in the layers where they believe they have the best shot of thriving. As with the beavers in Canada and bats in Texas, mole-rats and other animals select places that are food-rich, climatically comfortable, and safe from perceived threats. Stratifying vertically is a helpful way to avoid needless competition and overexploitation of finite resources, as demonstrated by the fact that even as mole-rats chomp away on the fleshy roots of wild sweet potatoes and cassava, those plants' leaves can be simultaneously enjoyed by antelope and other surface-level herbivores.

Earth's soil is much more densely populated than we may think. Across North America, groundhogs spend the majority of their lives tucked underground in multiroom settlements, sometimes dozens of feet long, where they sleep, hoard food procured above the surface, and hide away from hungry foxes and coyotes.

All around them, the dirt walls quietly creak and crumble as neighborly earthworms wriggle to and fro, feasting on an abundant supply of decomposing organisms. Among the earthworms, various types will occupy specific layers of dirt to more efficiently divvy up available resources and reduce conflict. In total, soils are home to more than half of all life on Earth, including at least 85 percent of all plants and 90 percent of fungi, making them a popular destination for all manner of subterranean creatures.

Animals often choose to inhabit certain layers to minimize their exposure to predators—but predators make decisions based, in part, on where prey is most abundant, creating a delicate dance. Soil may insulate worms from attacks by birds and amphibians, for instance, but no place offers complete security. Some worms learn this lesson the hard way when they shimmy through what appears to be a solid patch of soil, only to fall through the ceiling of an open tunnel dug by an insectivorous mole (not mole-rat), who swiftly bites the worm's head off before dragging the unlucky forager's body to a subterranean storage chamber.

In wet, mountainous areas like the Cascade Range of the Pacific Northwest, heavy snowfall several feet thick can create a temporary new layer, called the "subnivean" zone. Within this zone, mice, voles, and shrews ride out winter weather by carving insulated igloos between the soil and the bottom of the snowpack, where the temperature remains at a relatively balmy 32 degrees.

Similar stratification takes place up in the treetops. Kinkajous, for instance, are fruit lovers who inhabit dense rainforests, so their sweet spot is the lower canopy, or understory, which we now know they traverse on well-established arboreal highways. Life in the understory offers many clear advantages. It keeps them out of the way of the spider monkeys and coatis, who inhabit the tree layers above. It also minimizes their exposure to surface-level predators like jaguars and foxes, not to mention the eagles and hawks who lord down from treetop perches. For the kinkajous, their Goldilocks layer is just right.

In some cases, competitive populations will reach a mutual consensus on how to divide things. In tracking two distinct populations of warblers, for instance, researchers observed that one population lived entirely at an elevation below 4,800 feet, while the other was found only at 4,800 feet and above. In other words, instead of engaging in a daily food fight, the two populations apparently concluded that it would be easier for everyone if they just occupied different planes. Variations on these dynamics can be found everywhere from the tallest mountains to the deepest oceans.

As fellow members of the animal kingdom, many humans also weigh the pros and cons of life within various earthly layers. In South Papua, Indonesia, each hunter-gatherer clan of the Kombai people builds its own large tree house, where the male tribesmen sleep high above the mosquitoes and keep watch for encroaching enemies. (Female residents occupy small huts at the ground level.)

In the opal-mining town of Coober Pedy, Australia, where a hot desert climate can send summertime temperatures well into the triple digits, the majority of the town's 1,500 residents have taken the sensible approach of moving underground. Relatively soft sandstone has facilitated the construction of a wide array of subterranean structures, including not only personal residences—called dugouts—but also community spaces like churches, an art gallery, and a pub.

And in Cappadocia, Turkey, where the landscape is covered in volcanic rock called tuff—which, despite its name, is quite malleable—hundreds of underground cities have been uncovered, some of which date back to the eighth century BCE and descend eighteen stories deep. Within the region's rolling hills, tens of thousands of ancient peoples once stored goods, hid from foreign attackers, and lived year-round inside thermoregulated labyrinths just below the surface.

AFTER OBSERVING SEVERAL mole-rat clusters in various parts of Kasanka—some of which featured very recent elephant footprints, to the consternation of our armed scout—the rains became so intense that we had to scurry back to the park lodge before the roads turned impassable.

Sloshing out of Kasanka with my driver, a strong, silent type named Rodgers, and my local fixer, a twenty-five-year-old woman named Ketty, we turned onto the Great North Road, where Rodgers gunned it around the inner elbow of Zambia's flexing arm, entering the Copperbelt at 99 miles per hour.

The Copperbelt is the heart of Zambia's lucrative mining industry, as it was during British colonial rule, when the Northern Rhodesians and their partners took millions of tons of raw materials from the earth and put them up for sale on the international market. Today, it appears to be mostly Chinese companies that exploit this landlocked region for not only copper, but also limestone, coal, manganese, and gold. Open-pit strip mines visible from the highway seemed to go on for miles as double-wide dump trucks circled and excavators dug for treasure.

The other big business these days is selling carbon credits, which is what Ketty does for a living. For a certain price, she said, the Zambian government will agree to leave specific forested areas unscathed, or sometimes restore degraded tracts of land, so that oil producers and other multinational corporations can say that they're offsetting their enormous ecological impact by doing good elsewhere.

Passing the time en route to our next destination, I read to Ketty from a fascinating article in *National Geographic* about a recent study by University of Florida biologist Francis E. Putz. In the peer-reviewed study, Putz and his colleagues showed that pocket gophers in the southeastern United States excavate and maintain expansive tunnel systems up to 500 feet long, at great energetic expense. These tunnels aerate the soil and allow partially eaten roots poking through a gophers' dirt ceiling to take in more nutrients and regrow. The gophers' waste is also dispersed throughout the tunnels, which further fertilizes the soil, accelerating plant growth. Put simply, Putz asserts, the gophers are engaged in a rudimentary form of farming.

"Mole-rats do that," Ketty breezily interjected.

"Excuse me?" I said, noting that the study described the gophers as "the first farming non-human mammal."

"I worked with several researchers when they came to Zambia years ago, and they told me that some mole-rats do what you're describing. That the mole-rats would cultivate their food."

Intrigued, I WhatsApp'd two people: Dr. Radim Šumbera, the researcher who'd led the historic mapping excavation in Ndola years earlier, and Dr. Kyle Finn in South Africa. Šumbera replied right away. "Yes, they leave partially eaten roots, which are backfilled with soil so the roots can regenerate." He offered to send a copy of an old paper where this was mentioned as a footnote. Finn added that in his fieldwork in the Kalahari, he saw firsthand evidence of such activities, where he would dig up plants that store nutrients underground in bulbs, tubers, and the like (known as geophytes in scientific circles), some of which had clearly been eaten and regenerated over and over again.

If anything, Finn said, the evidence is stronger in mole-rats than it is in the Florida gophers. Rather than just spreading fertilizer around, the act of backfilling tunnels after eating part of a large plant encourages faster regrowth. Over time, the success of this technique—likely inadvertent at first—may have become an evolved trait.

Foraging is still the mole-rats' primary method of food collection, and if a plant is small enough that they can finish it in one meal, they usually will. But mole-rats have remarkably strong spatial skills, powered by charged particles in their eyes that enable them to navigate using Earth's magnetic field as a compass. They know every inch of their current and former tunnels, Finn noted, so in the case of large plants that are too big to eat in one sitting or carry back to their storage area, they will sometimes eat a piece, backfill the tunnel, and return later for another bite. "It's an and/both situation."

Doing some mental math, I began to ponder the implications of this. Mole-rats are believed to have arrived in this part of Africa during the Pleistocene Epoch, which began an estimated 2 million years ago, and their evolutionary ancestors date back even further. Fossil evidence suggests that the southeastern pocket gopher dates back roughly 1.3 million years. It wasn't until a little over 10,000 years ago that humans did anything we currently classify as "farming."

In other words, it's possible that mole-rats were the first mammals to farm in Africa—or anywhere else.

As I contemplated this audacious scenario, Ketty called up the local contact at our next stop, in the Copperbelt town of Mpongwe. While I had been distracted by all my WhatsApp messaging, we had apparently turned into a more rural area a good distance away from the highway. Pulling up to a field, we were introduced to Bridget Mulebi, whose family owns this land.

Speaking to Ketty in a local tongue, Mulebi guided us to the part of her land where mole-rats were most active. There are hundreds of mole-rats living here, she said. But unlike in Kasanka, this was no lush forest. It was a family farm.

"The mole-rats came here to eat your crops?" Ketty presumed.

No, the farmer said. The mole-rats were here long before her family arrived. Their farm, which grows maize and soybeans, is a more recent development that they have built on top. The mole-rats, Mulebi said, don't much care for her maize, preferring tubers and root vegetables, so they mostly eat the wild ginger that was already growing in this spot.

Taken aback, Ketty clarified, "So you aren't trying to get rid of the mole-rats living under your farm?" No reason to, Mulebi replied through a stiff smile. They do their thing, she does hers.

Silently absorbing this conversation in the blazing sun, I suddenly realized that a surreal spectacle was materializing before me: a farm plot that was being simultaneously harvested by two different families. One aboveground and one below.

When I later described this interaction to Dr. Putz, the University of Florida researcher who authored the pocket-gopher study, he grinned broadly. Hearing what African mole-rats have been up to, he said that calling it *farming* is "spot-on."

People will quibble about terminology, Putz said. When his research first appeared online, the Florida Farm Bureau reiterated its firm belief that agriculture is limited to "the science and art of production of plants and animals *useful to humans*"—a definition he views as arbitrarily narrow and self-serving.

Some people also claim that proper farming requires intentionally planting seeds before harvesting, but history is filled with examples of people cultivating food from perennial plants they did not originally put into the ground. "I think the whole issue is intellectually exciting," he said, "because it's not really settled."

At the time of our call, Putz was busy working in Australia, where he noted many Aboriginal people also have a long history of unconventional farming in ways that early European colonists stubbornly refused to recognize. Instead of sowing seeds in tidy rows, many Native peoples carefully removed plants that were competitive with the ones they preferred, to consciously but subtly encourage growth among their favorite foods. "To naive outsiders, they looked like hunter-gatherers, simply because the landscapes they harvested from were left wild."

OUTSIDE OF MAMMALS, several species have been shown to maintain their own food supply in one way or another, including leafcutter ants and some termites, who cultivate fungi to feed themselves or their young. One researcher has found evidence that some snails farm too.

Japanese scientists discovered that a particular kind of damselfish known as *Stegastes nigricans*, aka the "dusky farmerfish," uses advanced agricultural practices to farm their favorite kind of red algae, a rare delicacy known as *Polysiphonia* sp. 1. They do so by meticulously clearing debris and weeding out less palatable foods that might slow production of the desired algae, while fiercely defending their farms against thieves and intruders. As a result, they have constructed tiny agrarian settlements that provide a steady supply of their favorite food.

I once had the pleasure of seeing this cultivation firsthand while snorkeling in a secluded part of Ōdo Beach on the southern coast of Okinawa. To get there from the capital city, Naha, I took a series of local bus connections that zigzagged across the island before dropping me off about a twenty-minute walk from the shore. The final stroll to the beach weaved through a quilt of tidy, smallholder farm plots filled with sugarcane and winter vegetables. Looking at Google Earth on my phone, I was struck by how vivid the patches of various crops appeared—and the clear human influence their presence inferred. Were it not for us, these colorful rectangles would not exist.

When I reached the beach, I met up with a local dive instructor who guided me out to see the damselfish gardens in the middle of a lagoon. As we approached, I was encouraged to keep a respectful distance. Despite being only a few inches long, damselfish will not hesitate to attack humans if they believe their harvest is under threat, he said.

Levitating horizontally about five feet away from our first stop, the instructor raised his hand as if to say, "Go no farther." Looking down, I glimpsed the first of a string of tiny farm plots, hidden just beneath the waves. It was instantly recognizable because, true to form, the reddish-brown algae growing in it looked completely different from the surrounding coral reef.

From the surface, I watched as a single black damselfish paced back and forth like a night watchman, glowering at passersby and shooing away anyone who veered too close.*

Are these damselfish the only ones who do anything like this? Unlikely. By some estimates, barely 10 percent of ocean life is even known to us. Since most deep-sea worlds are off-limits to humans, at least for now, we don't yet have a comprehensive understanding

* In a curious twist, a different study, published in the journal *Nature Communications*, reported that despite their vigorous gatekeeping, some farming damselfish do allow one particular species to wander into their plots—a kind of shrimp whose presence (and fertilizer) the damselfish have learned helps their algae grow faster. The shrimp have since become dependent on the aggressive protection of damselfish, whose farms serve as a reliable safe haven from predators. The research team that discovered this pointed to it as evidence that the damselfish have in effect "domesticated" the shrimp over time, similar to how some wolves became the animal companions we now know as dogs.

of who lives where, much less how they interact with the resources around them. But even if we did set eyes on all the creatures of the deep, history has shown it's a dicey proposition to declare "Humans are the only species that . . ."

BACK IN MULEBI's maize field, Ketty reminded me that not all of Africa's mole-rats are as fortunate as those here or in Kasanka. Since the Stone Age, mole-rats have been more than just neighbors and nuisances to local villagers. To the undiscerning diner, they are a sought-after piece of game meat, obtained not with a spear, but through a honed practice of deception.

The most reliable technique, Mulebi's neighbor explained, is to use a garden hoe to chop a couple of feet into a fresh mound, exposing the mole-rats' tunnel to the open air. Just as beaver parents can instantly detect even the slightest change in water levels from inside their lodge on the other side of a pond, mole-rats can sense the introduction of an unwelcome breeze anywhere in their compound. A sign that somewhere, somehow, there's been a security breach. When a lumpy and skittish mole-rat inevitably rushes to rectify the situation, she'll listen carefully for the sound of snakes, jackals, or other potential predators who may be lingering nearby. Once she believes the coast is clear, she begins frantically plugging the hole with soil using her back feet. It's at this moment, the neighbor said, that he will ever so quietly raise his hoe above his head before slicing it down like a guillotine.

Zambians have a compelling case to say that it's arbitrary to draw an ethical distinction between eating a mole-rat and, say, a chicken. Even if local villagers have other protein options—never a certainty in these parts—a doomed mole-rat's final moments, unpleasant as they no doubt are, pale in comparison to the lifetime of unmitigated suffering in every buffalo wing. Still, in my heart, I couldn't shake the feeling that there's something despicable about slaying an animal for dutifully plugging a leak in their roof.

As we continued our walk through Mpongwe, I noticed for the first time that many of this area's mole-rat complexes appeared to be located near towering termite mounds more than a dozen feet high, giving open fields the appearance of graveyards. Through Ketty's interpretation, Mulebi explained that this is not by accident. In this part of Zambia, periodic flooding can create real problems for the mole-rats, who risk drowning in the waterlogged soil if they don't escape to higher ground. In times of crisis, they will sometimes climb up into a nearby termite mound or large anthill as a kind of life raft until conditions improve. Backup layers that allow families to ascend skyward when the going gets tough, before returning to what remains of home once the floodwaters subside.

In areas like Mpongwe, where mole-rat concentrations are high, researchers say it's also not uncommon for multiple tunnel networks to bump into each other underground, which can be resolved in any number of ways. If two solitary males, both living their best lives in hermetic isolation, happen to cross paths, things can get nasty. A fight could ensue and one or both will likely be injured or killed. Alternatively, knowing the potentially fatal stakes, the males could deem the risks of a duel too high and the payoff of stealing the opponent's tunnel too puny to take the chance. In such situations, the mole-rats will often seal up the mistakenly crossed path with a soil wall and quietly back away from each other like it never happened.

For social mole-rats, whose expansive networks are over a mile long in some cases, the odds of crossing paths with a neighbor are far greater. One researcher studied sixteen different African mole-rat tunnel networks and found that more than half of them connected to another. In one case, four separate compounds were linked together, collectively totaling more than four miles of tunnels.

In situations where open tunnels do connect, researchers do not believe this necessarily means that the families get along. The lack of physical barriers could indicate a standoff in which both sides covet their neighbor's property but are waiting for the other to die or move out. Less "keeping up with the Joneses" than burglarizing them. Mole-rats have a very powerful sense of smell, so individuals likely know by scent cues where their territory ends and their neighbor's begins, making wall construction unnecessary. What's keeping them apart isn't a physical barrier, but the mere presence of the other.

DOZING OFF IN the passenger seat during our long drive from Mpongwe to Mwinilunga, in the far northwestern corner of Zambia, I disappeared into a deep and lucid dream, a predictable side effect of the antimalarial drugs I was hopped up on. In my mind, I was sitting under an umbrella in a nameless public square, possibly somewhere in Europe, while passersby scurried under storefront awnings in the rain. Behind each of the people was a trail of GPS tracking lines, and the raindrops, I realized, were actually digital dots falling like hail from the sky. People were huddling together to avoid getting hit by the dots, while I watched motionless. After a few minutes of passive observation, I worked up the confidence to stand and braced myself to walk out into the rain. As I stood up, I bumped my head on the umbrella, which hurt more than I expected. Rubbing it, the umbrella suddenly hit me a second time, causing me to abruptly wake up, my head thumping against the car window.

The farther west we went, the worse the roads got and the bleaker the scenery seemed to become. A carousel of charcoal vendors, butchers selling raw chunks of fly-eaten meat, and rows of grass-roofed huts offering a seemingly identical display of watermelons, bananas, and tomatoes. Surrounding every village was a dense layer of discarded Styrofoam containers and plastic bottles. So thick was the detritus that one could distinguish between the newer bottles haphazardly tossed from a car window and the flat, tire-streaked bottles of yesteryear, pulverized so completely that they matched the soil.

Then, just as I started to close my eyes, eager to return to the village of dots and lines, I spotted something out the window and urged Rodgers to slow down.

"Are those mounds out there?"

Indeed they were. Just beyond the roadside trash barricade on the outskirts of Ndola were the remnants of dusty piles, indicative of mole-rats living below, either now or in the semi-recent past. Their presence here almost certainly pre-dated the town that had been built atop them.

This whole region has been transforming, Ketty said, as Zambia's human population has almost doubled in the past twenty years, putting enormous strain on local ecosystems. It was in the hills nearby where the Czech researchers first came to map mole-rat tunnels a decade and a half earlier, and the staggering complexes they uncovered were almost certainly not unique. At the time, their research took place in a small, forested area that was bursting with life. Today, the region is filling in with (human) housing developments.

By some estimates, humanity will construct the equivalent of a New York City every month from 2020 to 2060—a future currently previewed across sub-Saharan Africa. On several occasions, Ketty and I spotted what appeared to be a dry mole-rat mound in the distance, only to see once we got closer that it was a pile of a different sort—a pillar of concrete stabbed into the ground to support the building of a new structure.

Farther south, on the outskirts of what is now Lusaka, Zambia's chaotic capital city, small Ansell's mole-rats once built expansive tunnels of comparable length to those of the giant mole-rats of the north, despite having bodies one-fifth the size. But this admirable achievement has not spared them from facing eviction in one of the only places they still exist, the rapidly shrinking Lusaka East Forest Reserve.

Many Zambians I met spoke fondly of childhoods running and playing in what was a dense forest, now a shadow of its former self. Once 4,400 glorious acres—five times the size of Central Park—the reserve has been pared back in recent years to a mere 1,500, with plans to chop it down to fewer than 700 over the coming years to make room for new homes. Real estate advertisements emphasize that each new house comes complete with its own concrete security wall to keep intruders out.

On ostensibly government-protected land, local villagers have also taken to chopping the forest trees down to the roots to sell along the roadside as firewood. As the greenery has disappeared, it has triggered a death spiral in which less rainfall is absorbed into the soil, the land dries out, and much of the plant life withers away. Rivers that once flowed year-round have been replaced by dusty drainage canals that run only during a flash flood. A vibrant landscape slowly being replaced by the ruins of lost cities we never knew were there.

In such situations, the mole-rats' preferred method of building puts them at a distinct disadvantage. For us, the creation of a family home is generally marked through pouring concrete or laying bricks. We introduce something into the landscape. But mole-rats operate in a world of addition by subtraction, carving out room for life where there once was none. It's a brilliant model, but when our shovels slide down from above and the excavator buckets fill with dirt, the carefully constructed bedrooms and highways that gave their engineers a sense of place appear to our untrained eyes like empty space.

AFTER A BRUISING twelve-hour jaunt, we finally pulled into Mwinilunga, a remote town so ecologically different from those in the Copperbelt that it felt like another country. After hundreds of miles of tomatoes, this was pineapple territory.

Red-dirt roads struck a stark contrast with the lush greenery, baby-blue skies, and wispy clouds. To the north, the mighty Zambezi River rises from the woodlands near Ikelenge before snaking its way through a half dozen countries, slipping down Victoria Falls, and cascading into the Indian Ocean.

As recently as 2013, new kinds of mole-rats were still being uncovered in this fringe region, after eons of living under our radar. It's presumed that long ago, they splintered off from the giant mole-rats we saw in the Copperbelt, before evolving independently in this more verdant corner of Africa.

Meeting our host, a toothless man in a navy-blue work uniform named Reuben Chikombi, we were guided to his family's farm, located very far from what was already a lightly beaten path. After pulling up to his hut, we were soon joined by his son, Chibesa, a charming Gen Zer in an aloha shirt who guided us into a vast field of short green ginger plants. Interspersed between all the plants was a dizzying number of mounds. Hundreds of them. Possibly thousands.

"That's a bedroom over there. That's a road that leads to a sitting room," Chibesa said frenetically, rattling off his observations like a proud tour guide. "That's another home over there. They're having children, so they're building new rooms right now. . . . This tunnel goes from here to here, which connects this village to that one. . . . If they run out of food here, they will go over there and start a new village over there instead. . . . More houses over here. . . . This is a new one that just appeared after the rains started."

As my phone storage rapidly approached its limit, I desperately tried to jot down his comments in my notepad, but couldn't keep up. My heart started to race, knowing that I was missing bits and pieces of what he was saying. After ten minutes of fevered scribbling, I resigned myself to just soaking it in, accepting that I was never going to be able to capture all that I was seeing, and doing my best to visualize the hidden settlements he intuited from above. Ketty and I gave each other a knowing look, like "Holy hell, we've hit the mother lode."

The light color of the soil in the mounds is a useful indicator of what's happening underground, Chibesa noted, because drier soils suggest that it came from lower depths, where the most important chambers are placed. He's spent countless hours in these fields, trying to make sense of what's going on below his feet.

One emerging technology that could soon make such determinations significantly easier is ground-penetrating radar (GPR), which is beginning to be applied to animal worlds. Unlike excavation, which destroys a habitat in the process of exploring it, GPR sends out electromagnetic pulses that can reveal with high-resolution accuracy where gaps less than two inches wide can be found. Simply by pushing what looks like a fancy lawn mower over a particular area, researchers can survey the same spot over and over again, helping us to better understand who our subterranean neighbors are without upending their lives.

For now, most people still do it the old-fashioned way. Using his hoe, Chibesa opened up a handful of tunnels to demonstrate their varying sizes—signs that this field may be populated by both larger and smaller mole-rats.

Standing absolutely still, we only had to wait about fifteen minutes before the first hole came to life, as tiny paws started shoving mud around to plug the gap. Within another minute or two, no more light invaded their secluded world.

Getting back into our car, I asked this young man whether he had ever seen a map of what mole-rat tunnels actually look like underground. He had not, but said he'd like to.

As I pulled up the Ndola map on my laptop, his eyes grew large, and he yelped, "Yes! That is it!" Beaming at his dad, he exuded a sense of vindication that the hidden worlds he intuited from above really were there all along.

WHILE RODGERS HIGHTAILED it back on our multiday trek across the Copperbelt and onward to the international airport, I sat in the back seat and reflected on how wholly incomplete the view through my window was. Whooshing past village after village, I started paying closer attention to the gaps separating the brick houses and thatched stalls. The fleeting spaces between and underneath, where I scanned for iceberg tips in the dirt that suggested concealed kingdoms below.

As I did, I allowed my eyes to unfocus, giving way to the temporary illusion of double vision. Second huts, fainter than the originals, soon filled in what had been blank spaces. In the intervals between villages, where I had once seen only single-file lines of backpack-toting schoolchildren and tramping pedestrians, my exhausted head nodded off to the side, replacing this familiar montage with twin streams of passersby briskly shuffling atop and below the dirt. A hallucination that may, in fact, be closer to the truth.

Viewed on a flattened aerial map or a construction-site plan, such spaces, seemingly devoid of infrastructure, may appear at first glance to be vacant and primed for development. But as my time in Zambia vividly reminded me, the world we inhabit has fully three dimensions, with unseen layers that exist far beyond our perception.

Within those layers are infrastructures of a different sort. Destinations known and remembered. Strongholds to stockpile resources for leaner times. Shared spaces where new information is gained and exchanged. Sanctuaries where children grow up under their parents' watchful eyes. Transit hubs and high-speed routes relied upon to shuttle goods and community members to and fro.

As the afternoon drizzle started to fall again and the temperature precipitously dropped, my double vision was further distorted, refracted into multitudes through droplets on the glass. Suddenly, a scene once sparse was bustling and full. Everybody, it appeared, had someplace to be.

5

Safe Passage

Transit

AS WE SKIDDED off the main road and onto a mud path maintained only by tire treads, lines began to blur.

Flanking our two-track trail were a series of parallel lanes no more than a foot wide, which forked, merged, and crisscrossed our own. The dashboard GPS, relying on downloaded topographical maps, showed us headed toward a cluster of green and red dots. In the distance, I saw what looked like a ghost train, faintly visible and thrust forward by howling Siberian winds.

Through borrowed binoculars, I soon got my first glimpse of an ancient tableau, seemingly untouched since Neanderthals roamed the earth. On the far horizon, hundreds of nomadic travelers pierced the frigid air with their ringed horns, inhaling the brittle shards through Snuffleupagus snouts as they sprinted across the desolate steppe. Watching them charge through the treeless plain at full bore, I was reminded of a saying I'd heard shortly before arriving in this remote corner of Kazakhstan: "The saiga is a freedom-loving animal."

SINCE TIME IMMEMORIAL, saiga antelope have blazed trails across Central Asia while other civilizations rose and fell around them. They once inhabited a world dominated by woolly mammoths, saber-toothed cats, and giant deer with antlers 12 feet across—and outlived them all. When pioneering human caravans first attempted to navigate this unforgiving land, they considered the saigas to be their guides.

As autumn turned to winter, cold seasonal rains transformed once-green grasslands into sludge punctuated by tufts of withering brush. Driving through such muck is, I would soon discover, akin to controlled hydroplaning. The steering wheel bucked wildly as our tires swerved from side to side, shredding deep into the gunk and desperately gripping anything remotely solid enough to push off from.

Behind the wheel was Alexandr Putilin, one of two team members I was traveling with from the Association for the Conservation of Biodiversity of Kazakhstan, or ACBK. We were joined by Albert Salemgareyev, who was dutifully jotting down survey notes from the passenger seat. Between the two of them, they have more than twenty-five years of experience following saigas across the steppe, and I'd arranged to join them for the western leg of one of their winter field audits.

Our visit was timed to coincide with the rutting season, Albert noted, a critical part of the year for saigas. This is a gathering time, when antelope from across the region sprint together in massive single-file lines, before congregating for a series of competitive battles and subsequent mating rituals.

For males, it's a time for strutting and headbutting to demonstrate one's physical prowess to female onlookers—an activity so energy-intensive that many will forget to eat and keel over from extreme physical exhaustion. For females, this is a time of choosing. In addition to assessing fitness, they pay particular attention to one distinctive feature: a potential mate's floppy, elongated nose. The saigas' Star Wars–like sniffers are essential for filtering out dust in the semidesert plains and for warming frigid winter air before it reaches their lungs. A large nose is the ultimate sign of virility, so as the males marched into these annual gatherings, they were doing everything they could to accentuate their schnozzes, which swelled to twice their normal size.

Because of the saigas' preference for large-scale migrations across a landscape almost entirely devoid of trees, they have long been a favorite of global wildlife trackers. The isolation and starkness of the saiga world is a strength when it comes to GPS collars, which deliver a reliable stream of daily pings without risk of visual obstruction. For these reasons and others, saigas were among the first species to be monitored from space by the Max Planck Institute's ICARUS project. Indeed, even without trackers, many saiga congregations are large enough to be visible on ultrahigh-resolution satellite imagery, which has allowed AI researchers in Russia to scan for signs of where the antelope gather. Prior to such breakthroughs, conservationists would sometimes spend weeks roaming the boundless steppe unsuccessfully trying to locate the nomadic herds.

Tracking saiga movements in the modern day also provides useful historical insights, ACBK representatives told me. It's akin to opening a time capsule that illustrates what large-scale animal migrations may have been like when giants still roamed the steppe.

Yet for all of our technological advancement, there are still many knowledge gaps to fill, Albert said. The Ashiozek State Nature Sanctuary we were traveling through was created only a year prior, with its boundaries informed by GPS data that he and his team had personally collected. He seemed pleased that the animals were continuing to use the newly designated space as intended. Gazing out the window, Alexandr spotted another stream of the famously skittish ungulates wrapping around the contours of the earth.

What GPS coordinates do not yet tell us, Albert added, is what makes certain locations more appealing than others. Overlooking a nearby watering hole that the data suggested was a significant saiga stopping point, he asked himself: Why congregate here instead of at one of several neighboring ponds? Finding the answer requires some on-the-ground detective work.

Albert checked the camouflaged camera traps he'd installed on the adjacent trees, to try to make sense of what exactly draws the saigas to this body of water. He then bent down to collect some pH samples, which he uses not only to check the quality of the water, but also to analyze for environmental DNA, to see what other kinds of animals have been drinking from it. Puzzle pieces that can help illustrate the larger picture. As he did, wind ripped across the lake so ferociously that it created white-capped waves, and snow started blowing in sideways—the outer ring of an encroaching winter storm.

These were no fluffy flakes. This was the harsh, crunchy ice of a walk-in freezer, which instantly hardened our truck's caked-on mud like liquid nitrogen.

Seemingly unperturbed by the dropping barometer, Albert chirped, "Who's hungry?" Setting up an impromptu picnic outdoors, he and Alexandr dug into a prepared meal of fresh bread, tea, and dried meats of unknown provenance. In a muted attempt to thaw my face, I opted to guzzle spicy instant noodles.

A TRADITION OF nomadism has loomed large in this corner of the world since at least the Early Iron Age. The very word *kazakh* is thought to be derived from the Turkic word *qaz*, which means "wanderer" or "vagabond." For millennia, small human groups routinely migrated to higher altitudes in the hot summer months, where melting snow made for abundant water and greenery, before returning to sheltered valleys at lower elevations ahead of winter's whipping winds. At various times of the year, disparate nomads would also congregate for formal meetings and large cultural festivals, complete with physical challenges like wrestling and tug-of-war. When the gatherings were finished, the nomads dispersed once again.

Historical analysis has shown that nomads' seasonal migrations and time-honed transit corridors may have helped pave the way for portions of the Great Silk Road, which was less a singular route than an ever-evolving web of pathways and bypasses that linked together reliable stopovers. At the Silk Road's peak, this corner of western Kazakhstan along the fertile banks of the Ural River became a vital transcontinental crossroads. To this day, the river still serves as a natural boundary between Asia and Europe, meaning this part of Kazakhstan is, quite literally, where East meets West.

The land above the Caspian Sea also served as a main artery that linked together vital transit points. From Zhaiyk in the north to Saraychik in the south, strategic riverbank settlements allowed weary caravans to rest their heads, unwind in large public baths, exchange knowledge, stock up on food, water, and other essential supplies, and regroup before moving onward. Saraychik was also a key intersection between Central Asia's north–south and east–west corridors. From there, one could go anywhere.

Among overland voyagers traversing the continental divide, saigas have long been spoken of with an almost mythical reverence. Cave paintings of the antelope created as far back as the seventh century BCE depict them as resilient, elusive, and charismatic. Expert navigators who embody the true spirit of a most inhospitable terrain. Even now, saigas are proudly depicted on Kazakhstan's national currency, racing toward greener pastures on the far side of the 2,000-tenge bill.

In human society, traditional nomadic practices persisted in this part of Central Asia even as the land itself was variously claimed and reclaimed by Turkic tribes, Genghis Khan, and the Russian Empire. Only with the rise of the Soviet Union in the twentieth century did mandatory sedentarization efforts successfully manage to lock nomadic people in place, force them to farm, and extinguish a cultural flame that had burned brightly for millennia. Denied the ability to live in perpetual motion, people's navigational muscles began to atrophy.

This transition was viewed by many as an inevitable shift toward "modernization," but it also represented a significant diversion from our own history as a species. Of the roughly 300,000 years that *Homo sapiens* have existed, we spent the first 288,000 of them living in small, mobile groups that followed the migratory patterns of other animals. Forcing nomads into agricultural plots was also a colossal failure because much of the steppe soil was unsuitable for growing crops, which is why so many people were nomadic in the first place. Iconic filmmaker Werner Herzog once remarked that "whatever went wrong and makes our civilization something doomed is the departure from the nomadic life."

Whether by choice or necessity, most Kazakh people now lead stationary lives in urban centers. For many, the timeworn landmarks and pathways their ancestors depended on have faded into insignificance.

The world that nomads once freely roamed has also been feverishly transformed, not only by agriculture but through the construction of new roads and railways. When the Soviet Union collapsed, an additional series of barbed-wire border fences drew iron lines across the earth, erecting man-made barriers on top of natural ones.

For the saigas, whose nomadic days never ceased, these new layers now represent an existential threat.

WITH THE WEATHER taking a turn for the worse, Albert and Alexandr eventually decided to call it a day and start fresh in the morning. After a blistering drive through whiteout conditions, we hunkered down in a motel in the tiny village of Kaztalovka, less than 10 miles from the Russian border. Albert pulled out his laptop and showed me some of his tracking data to demonstrate how saiga movements have changed over the years.

"All these saigas were once part of a single massive population group," he said. "They would routinely trek hundreds of miles as part of an expansive migratory route across the Eurasian steppe. Saigas could be found from the Caucasus Mountains to Mongolia, and everywhere in between."

For millennia, they lived with few barriers and obstructions on the landscape. Then, seemingly overnight, rattling railways and roaring roads began to crisscross the region, fragmenting the population into smaller, more isolated groups and cutting whole populations off from their water and food supplies. As the interior steppe got more accessible to vehicles and passengers, development begat more development. When the Soviet Union collapsed, nation-state borders further restricted saiga movements.

On his screen, Albert clicked open a recent map that showed GPS-collared saigas walking right up to what is now the Russian border and looking frantically for a place to cross. In their bones, they felt they must get to the other side, just as their ancestors always had.

To the south on the Ustyurt Plateau, the data showed another large group that once routinely traveled deep into what is now Uzbekistan before being unceremoniously clipped off by a new rail line in 2014. The data from the years prior to the railway's construction showed a flurry of squiggly lines representing individual animals

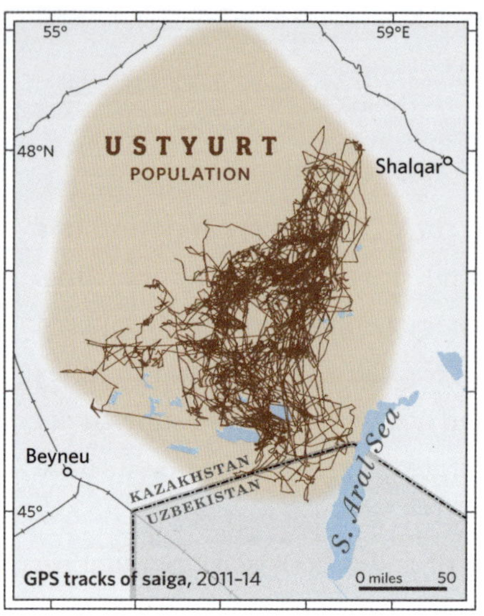

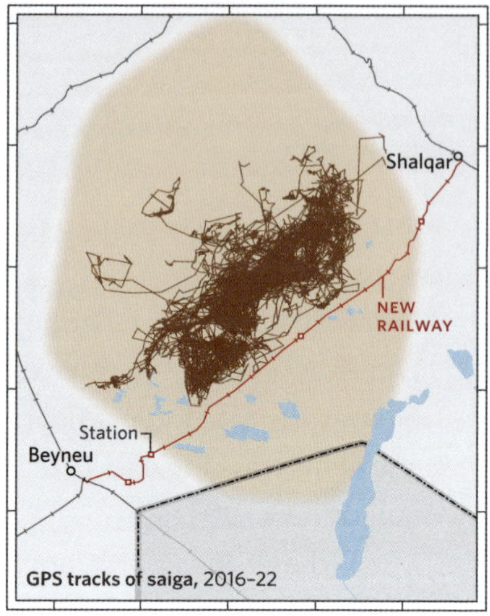

zipping back and forth with the seasons. The post-construction slide showed those same animals spinning around despondently on the northern side of the tracks, while the fresh grasses and water they had come to rely on withered away to the south.

For an individual who was born to run, it's hard to imagine a worse fate.

One might reasonably ask, as I did: Why don't the saigas just walk over the tracks?

But we must remember, Albert said, that saigas have spent nearly every waking day since the last ice age roaming vast open spaces. In a region where hiding places are few and far between, the best defense is to flee before a threat gets anywhere near striking distance. To survive, their large communities have developed a hyperalertness to the faintest sight of encroaching predators. ICARUS data has revealed that when saigas choose a place to give birth, they will intentionally pick spots that straddle the known ranges of local wolves to minimize the risk of encountering one.

It's an evasive strategy that served the saigas well, at least until the stillness of the steppe was brusquely pierced by the arrival of mysterious steel snakes, clanging and screeching as they careened across the landscape. In that moment, most saigas appear to have updated their mental maps and cordoned off whole sections of their former range as simply too dangerous for any self-preserving antelope to consider approaching.

The damage from this disruption was twofold. In the short term, it led to an unanticipated shortage of food, water, and shelter from the elements, akin to completing an arduous pilgrimage to a trusted oasis only to find that while you were away, it had become a sand-blown mirage. Disbelief turned to panic, followed quickly by desperation and starvation.

Long term, such interruptions can lead to a complete collapse in spatial knowledge—a loss that science writer Ben Goldfarb once wrote is "as thorough as the erasure of a language." Just as human nomads forced to stay in one place eventually lost connection to the physical landmarks of their ancestors' journeys, young saigas learn where they can reliably find food and shelter, as well as places to stay away from, through active participation and observation.

This cultural exchange keeps their entire world in orbit and is one of the reasons they have been successful at navigating a terrain fatal to so many before them. As the land they know is continuously cleaved and partitioned, whole stretches of the Saiga Silk Road begin to fade from memory.

THE NEXT MORNING, I awoke at 4 a.m. to an unusual buzzing sound emanating from the bathroom. In the pitch-black of my motel room, I saw a neon rectangle glowing erratically on the far wall, which I soon realized was pulsating in sync with the gusting wind outside. Traveling through what is now Kazakhstan in the 1800s, British adventurer Frederick Burnaby feared this wind the most, writing,

"It blows on uninterrupted over a vast snow and salt-covered tract. It absorbs saline matter, and cuts the faces of those exposed to its gusts. The sensation is like the application of the edge of a razor." I glanced out my picture window to see dim streetlamps flickering on and off, momentarily illuminating a ferocious blizzard before returning to complete darkness. Between squalls, it was eerily silent.

Winter in the steppe is, I would learn, no place for a morning person. The predawn hours grind by at a glacial pace and the sun remains completely absent until almost 9 a.m. Wizened locals appeared to cope by entering a semi-hibernation, sleeping in late and retreating home early to what must be some of the world's most well-insulated buildings. Fortified with double-layered doors and fed a ceaseless current of hot air through a lattice of elevated gas pipelines a dozen feet above the frostbitten earth, small-town dwellings glowed like oil lamps. The grand payoff came only when daylight did finally start to crack through, briefly saturating the snow and sky in the most radiant shade of pale blue.

Over a breakfast of pillowy potato-and-cabbage piroshki, I was informed that the road we were driving on yesterday was now closed to vehicles due to the storm covering it in thick layers of treacherous black ice.

That sounds like a problem, I told Albert.

"No, we'll just leave by driving through the fields, where there's nobody to stop us," he replied.

Rolling out of town, I was astonished by how austere the steppe looked now compared with the day before. Gone was the slippery muck and browning grasses of shifting seasons, replaced by a blanket of crunchy snow. Against this stark black-and-white backdrop, the fresh crystals functioned like fingerprint dust, illuminating the only remaining pops of color: the rust-red streaks of human and nonhuman pathways in the dirt.

Parked in an otherwise empty tract of prairie land, we watched in stunned silence as thousands of saigas seized the day, flying like banshees across the periphery, their hooves gouging the frozen crust and stamping pigment into an earthen canvas.

WHETHER CHARTING A course between vital resources or cutting through jagged mountain passes, safe and efficient transit networks are a central fixture of many of the world's great civilizations.

Forest elephants have been shown to forge extensive networks of semipermanent pathways through the undergrowth of the Congo Basin, which connect watering holes, resting areas, and abundant food sources. Knowledge of these intricate routes is passed down intergenerationally, with the oldest matriarchs of the group presumed to possess the most comprehensive mental map of all. Elephant trails have such a profound imprint on the landscape that if foot traffic is heavy enough, over time it can compress the soil and create canals that link isolated rivers or ponds together, potentially opening the floodgates for aquatic animals to forge new pathways into uncharted waters.

Arctic caribou also create and maintain extensive trail systems that link together their summer feeding areas, winter shelters, predator-free calving grounds, and strategic crossings through valleys and rushing rivers. Given their constantly tramping through the snow and vegetation, their heavily trafficked corridors can last for centuries.

Research published in the journal *Science* shows that among some ungulates, such as the bighorn sheep of the Rocky Mountains, knowledge sharing and cultural transmission are key attributes of maintaining such routes, helping young travelers locate reliable, high-quality foraging spots—a process researchers call "surfing the green wave." In situations where information exchanges are interrupted by new obstacles or mass die-offs, it can "expunge generations of knowledge." To preserve such transit networks, the United

Nations released its first Atlas of Ungulate Migration in 2024—a living map that uses GPS tracking data to identify high-traffic corridors relied on by saigas, mule deer, ibex, gazelles, reindeer, zebras, and other hoofed nomads.

Whether made of dirt, rock, or concrete, transit systems matter because they signal that the journey ahead is well trodden. And no matter who clears the way, once a path exists, it quickly becomes a public utility.

Daniel Boone's famed Wilderness Road, which plodded a westward path through the Cumberland Gap and into the frontier beyond the Appalachians, was based on the Warriors' Path, a network of Native American trails that facilitated travel, trade, and communication between various tribes. But the Warriors' Path was itself built on trails blazed by bison and other large mammals who had already identified the paths of least resistance, from shallow river crossings to natural valleys.

The same could be said for the rim-to-river paths of the Grand Canyon, where modern-day tourists descend 5,000 feet down on trails used by Native Americans but blazed by bighorn sheep and mule deer.

In Australia, kangaroos often follow distinct navigational tracks between familiar locations, which some Aboriginal peoples have enshrined into the public memory through traditional oral storytelling, passed down through the generations.

Conversely, once humans built the first marked routes designed for automobiles, initially known as "auto trails," pigeons and bees began using them as visual aids for aerial navigation.* Baboon troops

* Bee navigation is the subject of much scientific debate, as their lack of hippocampi and other advanced brain structures would suggest limited mental capabilities. It's been speculated that to overcome this, bees use "navigation memory," imprinting a map onto their brains and hopscotching between linear landmarks, like pilots in the early days of aviation.

have also been shown to follow human roads when they need to get somewhere quickly. White-tailed deer, like those found throughout American suburbs, are perfectly capable of maintaining their own backwoods trails to follow year after year; but if humans decide to clear a nice, wide path through the forest, the deer are happy to conserve some energy and take ours instead. A good route is a good route.

Oftentimes, such interspecies exchanges happen completely outside of our comprehension. As author Robert Moor wrote in his superb book *On Trails*, "Humans are neither the earth's original nor its foremost trailblazers."

In making his case, Moor investigated a fascinating situation in Botswana's Okavango Delta, where a large number of zebras mysteriously disappeared at the beginning of the rainy season. Since conventional wisdom was that this particular zebra population spent their entire lives in the delta, it was presumed that they must have been eaten by lions. So you can imagine the researchers' surprise when GPS-collar data showed that not only were the zebras alive, but they had somehow trekked halfway across the country to a salt pan where fresh grasses were starting to sprout.

The plot thickened from there. After digging into old hunters' and explorers' records, researchers discovered that the migratory route the animals now appeared to be following was in fact one that their zebra ancestors had also used prior to 1968. That was the year the Botswana government installed new fencing that disrupted the zebras' migration, as it did for many other animals. Over the next thirty-six years, generations of zebras came and went before the fencing was finally taken down in 2004. As a result, no modern-day zebras ever had the opportunity to use their ancestors' route. This raised a new question: How did they know where to go?

The likely answer, it turns out, was elephants. In the wild, African elephants typically live for sixty to seventy years, meaning that, unlike the zebras, many local pachyderms *did* have firsthand knowledge of how to get from the delta to the salt pan. When the fencing was removed, the elephants apparently saw an opportunity to repave the migratory road they remembered traveling on almost four decades earlier. The zebras then saw the elephants forging what appeared to be an efficient route to abundant food and water, and decided to follow along. In the process, the zebras unintentionally retraced their distant relatives' footsteps and a long-lost trail was blazed anew.

WHILE THERE ARE clear advantages for the saigas in returning to familiar parts of the steppe year after year, the predictability of their movements has also been exploited at times by those who wish them harm.

When a team of scientists in the 1970s flew over a remote corner of the Ustyurt Plateau near the modern-day border between Kazakhstan and Uzbekistan, they spotted what appeared to be large undulations in the landscape that resembled miles-long arrowheads, with outer walls more than 10 feet tall and wide. An additional 20-foot-wide ditch along the interior side made the walls appear even taller from within.

What at first glance could be mistaken for remnants of long-abandoned fortresses, Silk Road settlements, or religious complexes were, in fact, structures built with a more nefarious intent: They were traps.

Known as *arans*, these devices were custom-made by ancient Turkic people for the expressed purpose of capturing sprinting saigas by the tens of thousands.

The arans' design is as clever as it is wicked. Understanding that the saigas would routinely migrate across the plateau through certain natural corridors, hunters manually reshaped the land to slope downward so subtly that it was imperceptible to those viewing it from the ground level. As it delicately descended, the earth would funnel the sprinters into a huge basin at the tip of the arrowhead, where they would eventually run into a deep ditch and steep wall. Having no ability to continue straight ahead, the saigas would instead dart to the left or right, seeking a way around the wall. As they did, they were unknowingly guided through ever narrower openings before reaching a dead end.

By this point, the saigas would realize that they needed to backtrack and find an alternative route. It was in this moment of confusion and reorganization that looming attackers, who had been quietly trailing the antelope all along, would make their move, mass-exterminating whole saiga caravans like fish in a barrel.

Among anthropologists, the Ustyurt arans have been hailed as

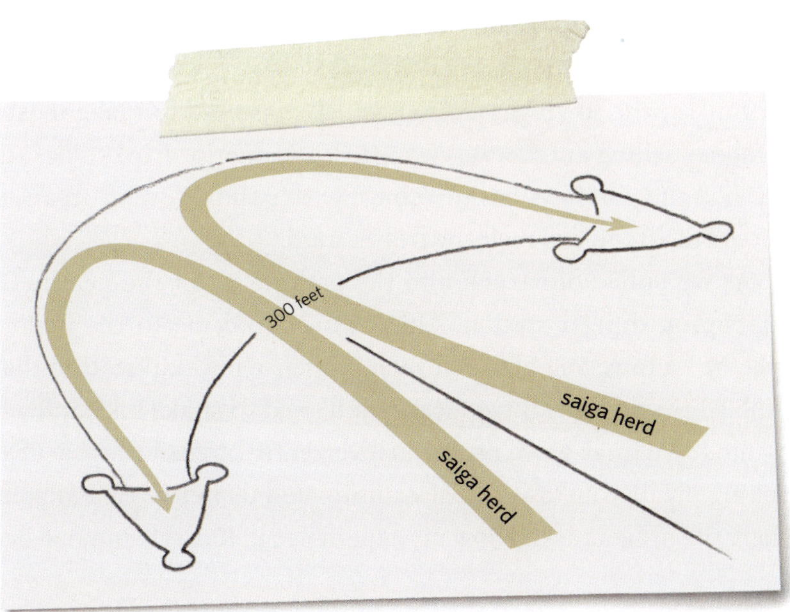

exceptional examples of human ingenuity. Many of the structures still stand to this day, as Albert showed me on satellite imagery. In 2021, Kazakhstan submitted a UNESCO proposal to designate them as being of "outstanding universal value."

The arans' continued existence on the landscape so long after their builders moved on is emblematic in a way of how, between the dividing lines of newer infrastructure, some corners of the saigas' ancient world do look much as they did thousands of years ago.

Standing in the middle of a frozen field while Albert and Alexandr were off analyzing the pH of another lake, I couldn't help feeling an acute sense, however deceptive, of timelessness.

BY OUR THIRD day in the steppe, the temperature plummeted below zero. Back home in LA, it was a sunny Thanksgiving Day, yet there I was holding up my phone on a family Zoom call, bundled in all five clean layers I had left. "Did you see the saiga?" relatives asked. "Are you getting everything you wanted?"

After another midday breakfast of cabbage piroshki, we filled up our tank at the town gas station, which appeared to make most of its money selling antifreeze, cigarettes, and energy drinks. Alexandr splashed the windows of our vehicle with scalding water in an unsuccessful attempt to melt the frozen mud.

As we pulled our truck into the parking lot of the nearby Okhotzooprom ranger station, Albert introduced me to the regional director—a towering figure of such immense physical stature that I fumbled through my introduction and tried to shake his hand with my muddy winter glove on, an inadvertently offensive move. Overlooking my discourtesy, the director invited us in for tea and bread, where the rangers asked how my experience in Kazakhstan had been so far. I said something about how the people had been very welcoming and drew a half-formed metaphor between the hostile weather outdoors and the warm atmosphere inside.

Albert said aloud that it was a customary sign of respect in this region to offer special guests a platter of sheep's-head cheese, which is, in fact, not cheese, but boiled meat-jelly scooped out of an animal's skull, served cold. Sensing that I would not want to partake in this ritual, Albert informed our hosts that they should save the delight for themselves. The director then pointed at another heavyset staff member and said, "He's vegetarian too—except that he eats people!" Everyone roared in laughter.

As we headed out into the steppe alongside the ranger convoy, Albert explained that one of their primary responsibilities is to root out poachers, for whom bagging a saiga had become a desirable prize. While saiga meat is considered a delicacy among local people, the real money is made by selling the horns, which traditional Chinese medicine practitioners cut into shavings or grind into a fine powder on the premise that its consumption can solve everything from a sore throat to liver issues. Such assertions largely crumble under even the slightest scientific scrutiny, and you don't have to be a pharmacologist to know that horns made almost entirely of keratin—the same substance found in human fingernails—are no miracle cure.

Still, such superstitions created an enormous market demand and drove the saiga into "critically endangered" status. So long as population numbers remain below a certain level, hunting is strictly prohibited, and it's the rangers' job to keep modern-day caravan plunderers at bay for long enough that saiga numbers can climb above the endangered threshold. At least one local ranger had already been killed for getting in the poachers' way.

Population figures are tabulated on a regional basis, and ACBK's aerial estimates show that this saiga group in western Kazakhstan may be the first to have surged back above that line. Albert's estimates showed that the western population had soared from only 135,000 five years ago to more than a million today, which he believes the rangers deserve the lion's share of the credit for.

Once saiga numbers become large enough, though, strict hunting restrictions will begin to fall away. At that point, a certain percentage could again be captured with giant nets, shot, and have their horns sawed off for sale on the international market. It's an open secret that this is the long-term payday that Kazakhstan is holding out for, which has helped fuel their conservation efforts.

Cruising on an ice road alongside the rangers, we used walkie-talkies to draw eyes to the various saiga convoys in the distance. By this point, my vision was better trained to know what to look for, but I still took a certain amount of pride in being the first of the team to spot a caravan on the horizon. Albert estimated the size of one group we passed at being more than 10,000.

As we rolled along, I then spotted what appeared to be large lumps of snow along the roadside, popping out among otherwise flat plains. Only when we got closer could I see that they were, in fact, scattered remnants of saiga bodies, picked clean by voracious eagles. As we got out of our vehicle to take a closer look, I remarked to Albert that they must have all been female, as I saw no visible horns. "No," he said. "The horns have been collected. These animals were shot."

Whipping out my phone, I began to take pictures of the gruesome scene. As I clicked away, my bare hands went numb in the bitter cold while the rangers looked on uncomfortably. After I captured the story from every conceivable angle, I walked back toward our truck, only to be approached by one of the rangers. "I need to ask you to delete those photos," she said firmly. Taken aback, I suddenly realized that perhaps we were not all viewing this discovery through the same lens.

When we got back in our vehicle, I asked Albert whether the carnage we had seen was indeed the work of poachers or whether this was perhaps the start of a government-sanctioned hunt, from which horns were being stockpiled for future sale. Staring stoically out the window, he met my inquiry with silence and pursed lips, as if to say, "Better not to ask."

I would later learn that indeed a regional hunt had been authorized a few months prior, following complaints from nearby crop farmers whose cultivated plots now dotted the landscape. The rebounded saiga populations were a nuisance, the farmers said, and "adversely affect the rural economy." It's a familiar refrain I'd heard many times in my food-sustainability work, where farmers will fundamentally change land use in a way that upends the lives of wild animals living nearby. Then, they blame the wildlife for causing trouble.

As part of the authorization order, special teams had been established with a target of wiping out 200,000 saigas from this western part of the country, just as their numbers had begun to recover. In a cruel twist of fate, the hierarchy was such that some of the responsibility for carrying out this purge fell to the same anti-poaching rangers who have dedicated their lives to driving out plunderers.

I pondered these conflicting winds of progress and setbacks as we turned onto the main highway. Alexandr cranked up the '90s Russian pop music on the radio, and we raced to get back to Kaztalovka before the sun disappeared for the day. I used my hand to wipe the steam from the inside of the window to get a clearer look at the vast plains beyond. Upon doing so, I spotted a massive single-file line of saigas running alongside us. When I pulled out my binoculars

to get a closer look, I was surprised to notice that the farther up the chain I scanned, the larger the animals appeared. As I turned my head back toward the road, Alexandr abruptly hit the brakes and the cargo-truck driver to our left aggressively honked his horn while the saigas boldly cut right in front of us all.

Streaming across the frozen asphalt by the hundreds, the antelope skidded frantically for a solid two minutes before disappearing into the endless, darkening steppe.

BIDDING FAREWELL TO Albert and Alexandr the next morning, I hitched a ride with a group of rangers and government officials

who were caravanning up to the international airport to pick up a BBC nature-documentary crew. I was heading northeast with the intention of then immediately flying south to the Caspian Sea port town of Atyrau. From there, I had a train ticket, departing in forty-eight hours, which would zigzag across the country before terminating in Almaty, Kazakhstan's largest city, more than 1,500 miles to the east.

I had arranged to take this circuitous route partly because I had been warned that driving across the steppe in wintertime is both dangerous and unpredictable. Travel delays in remote areas are often counted in days or weeks, requiring a level of patience that does not come naturally to me.

Due to high winds, the road conditions on the way north were rough, but manageable. What should have been a four-hour drive was expected to take eight, but my flight wasn't until the next morning, so I was not concerned. That is, until the call came in. The BBC crew's flight had been canceled. In fact, the airport had, I learned, been closed for more than four days. Unbeknownst to me, my flight into town earlier that week had been one of the last cleared to land before they shut it down. None of the daily flights to Atyrau since then had taken off.

Having traveled more than halfway to our destination, my escorts decided that they would hang out near the airport until flights eventually resumed. But my itinerary was suddenly much less certain. Online, my flight the next morning was still listed as "scheduled," but it was not at all clear that we'd actually leave the ground.

After catching a few hours' sleep at an Airbnb in a cold, brutalist condo tower, I woke up early and hailed a cab to the airport. On the drive there, the pea-soup fog was so thick that we could barely see 20 feet in front of us. When I walked into the airport lobby, I was greeted by dozens of grumpy passengers, some of whom were surrounded by several days' worth of empty coffee cups and concession-snack wrappers.

"Will the flight to Atyrau take off?" I asked the airline representative through my translation app. "It is scheduled," the agent replied emotionlessly. "Yes, but will it actually take off? I hear that it has not in several days," I pressed. "It is scheduled," she said again, this time with more force.

"This is what they do," a middle-aged man with a distinctly American accent said as I stepped aside. "In cases of cancellations, the airlines are on the hook to pay for customers' lodging and other accommodations, so they will just delay flights indefinitely to whittle down the number of people they ultimately need to pay for. Look outside," he said, pointing to the complete lack of visibility. "No way that plane takes off today."

The gentleman, who introduced himself as Chez, worked for an oil-supply distribution company. He was based in Atyrau but flew all over the steppe for business, so he'd seen this game play out many times before. "In winter, I've seen people stuck here for two weeks."

Sensing the way things were headed, Chez started to make alternative plans. As he stood next to a shelf of canned bear meat in the lobby gift shop, he called up his company's ground-transport service. The idea of driving across the steppe at night was a nonstarter, so if he was going to travel to Atyrau by road, he'd need to either leave soon or wait until the next day. As he weighed his options, I took the opportunity to pitch myself. "If you do go by car, do you think there might be room for one more?" He paused and smirked before saying, "Sure, why not."

Just then, the news came through the loudspeaker. Our flight was officially canceled. Chez seemed relieved to get that clarity rather than having the airline drag it out. If we left now, we could still get to Atyrau before nightfall.

He called back his transport service. "OK, let's go."

DRIVING SOUTH ALONG what was by now an all-too-familiar route, I looked longingly toward the vast continent to the east. With no saigas visible this close to the highway, I passed the time in the back seat counting stranded vehicles lodged in roadside ditches. I spotted three within the first two hours, all bald-tired sedans that had no business driving on ice.

When we finally pulled into Atyrau eight hours later, I checked into a rental apartment overlooking the Ural. Seated at my dining room table, I devoured the spiciest bowl of noodles I could find, the glorious steam from which fogged up my glasses, giving the river an ethereal smokiness. After a week of zigzagging across the frozen plains, this face-melting meal was pure ecstasy.

Rising early the next morning, I walked south along the riverside and turned left at a quaint gazebo simply labeled EUROPE. A five-minute stroll over the bridge to the eastern bank led to a matching gazebo, aptly labeled ASIA. Just like that, I had crossed the continental divide.

With about half a day to kill before my evening train, I hired a driver to take me 30 miles outside the city to Saraychik, the ancient Silk Road settlement that once served as a grand crossroads between disparate human nations. One of the more surprising things I came to learn while in Kazakhstan is that many of the most famous Silk Road sites have almost no demarcations or protections from desecration. Such was the case at Saraychik, which we reached only after driving through a farmer's field and asking a local neighbor for directions. Twice.

When we finally located the right spot, there was no fencing surrounding the site, no ticket booth, no visitors. Barely any signage at all. What was there was an enormous grid of stone, grass, and mud walls, each containing diminished doorway remnants that rose just above my waist. Much of the site, my driver noted, has been slowly consumed by the Ural. In a few places, amateur archaeologists maintained open-air excavation sites, sorting recovered shards into two piles: one for clay wall fragments, the other for skeletal remains of a wide variety of ungulates. Artifacts of our collective history at the crossroads of the world.

Looking out over this earthen schematic, I tried to envision what it must have been like in the thirteenth century, when this was one of the biggest cities of the Golden Horde, once a division of the

Mongol Empire. In its heyday, this site boasted many *caravanserai*, or road inns, for rest and recovery among exhausted travelers, as well as a wide range of workshops specializing in everything from metallurgy to pottery, from which traders could buy and sell goods before continuing onward into the great unknown. Preservation of such settlements, however temporary, is essential for the continued survival of long-distance journeys. So, too, I have learned, is a willingness to adapt when transit disruptions force unexpected itinerary changes—something the saigas know all too well.

As I gazed east from the riverbank toward the colorless horizon, I pondered what it must have felt like for an intrepid merchant to know that they would not see another human being for many days, maybe weeks. To be fearful of what you may soon encounter, but take solace that the space between is not wholly empty. For there is life beyond humanity's margins, and among nonhuman communities, that in-between space is the center of everything. That, I imagined, must have offered some measure of assurance that the road ahead, perilous though it may be, is well traveled.

SETTLING INTO THE top bunk of my four-berth roomette, I looked at a map of the Kazakh rail network, which is itself a relic from a bygone time.

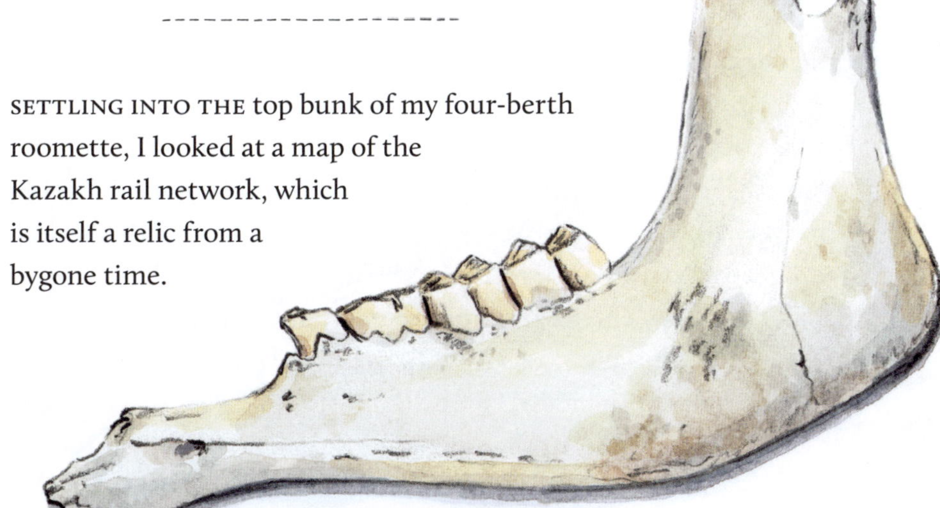

Like both the human and saiga Silk Roads, the railways pre-date modern borders, having been constructed largely by the Soviet Union when there was regular transit between various member states. When that system collapsed and the border lines were redrawn, many train stations suddenly found themselves on the wrong side of the tracks. The Oral–Almaty line would have taken me across the country to my desired destination—except that one of the line stations is located a smidge into southern Russia, where it makes a brief stop before immediately dipping back into Kazakhstan. That momentary detour is enough to make the line inaccessible to foreigners like me, who are required to obtain a Russian visa to board the train. Similar snags exist all throughout the Central Asian railway network, where half-empty railcars casually meander across nation-states as though trapped in amber from an earlier era.

To circumvent this obstacle, the route I selected took a highly inefficient shape that scribbled its way to the northeast, then to the far southeast, and then northeast again. This zigzagging infrastructure helped splinter what had been one expansive saiga nation into at least three fragmented pieces: the Ural population I was bidding farewell to, the Ustyurt population near the Aral Sea, and the Betpak-Dala population, which occupies a mammoth stretch of central Kazakhstan.

The train itself was spartan but functional, with a basic mattress, a stack of plastic-wrapped sheets, and, in the hallway, a Russian hot-water dispenser called a samovar. The sun was long gone, meaning that as with the terrain we were traveling through, the halls were shrouded in darkness, interrupted by light blips and the sound of clinking glasses emanating from roomettes. Cradled by the gentle sway of the thumping tracks, with visions of seasonal saiga settlements beyond my window, I fell asleep instantly.

Shuffling around in the middle of the night, I reached for my phone, assuming that it must be 2 or 3 a.m., given the stillness on board. Nope: 7:34 a.m. The next day was already underway, but you wouldn't know that by the look of things.

Restless, I ventured into the dimly lit hallway and stared out the window into the treeless void, which was intermittently basked in light each time we passed through a tiny human town. I could tell when we were entering one because my phone would briefly regain service as we approached, before it disappeared again as soon as we left. The township archipelago was remarkably consistent in its spacing, having once served as designated resting points for long-distance journeymen.

To the northeast lay the vast expanses of Siberia, where intrepid Russian scientist and geographer Prince Peter Kropotkin once undertook a five-year, 50,000-mile geological expedition. Along the way, he encountered a wide array of animal groups, including "ruminants whose social life and migrations offer the greatest interest," but "do not let man approach their herds." Through Kropotkin's keen observation of these secret ungulate societies, he gained a new admiration for their complex levels of cooperation. If we study animals, he wrote, not merely "in laboratories and museums" but "in the forest and the prairie, in the steppe and the mountains," we can see that their communities are not defined by "warfare and extermination," but "perhaps even more, of mutual support, mutual aid, and mutual defence."

Such cooperation, he noted, appeared to go beyond mere instinct, and mirrored community dynamics he'd also observed in the human villages that dapple the Siberian landscape. This realization that altruistic collaboration can exist in the societies of ruminants and human peasants alike, even in the absence of any formal government structures, helped turn the avowed socialist into one of the leading anarchist thinkers of the nineteenth century. Left to their

own devices, Kropotkin reasoned, humans living in complete freedom and autonomy would naturally cooperate for the greater good, just like the ruminants of the steppe.

I stared aimlessly out the window for about an hour before another soul appeared, a man in his thirties with short hair who emerged a few cabins down the way. He nodded politely and then stared out the window alongside me.

He started speaking in Russian but quickly deduced that I did not understand what he was saying. Reaching for my translation app, I asked him to repeat it. When I did, the app translated his words to "Do you like the Golden Horde?" Perplexed, I said I wasn't sure what he was asking. Trying again, he said, "Are you interested in the Golden Horde?"

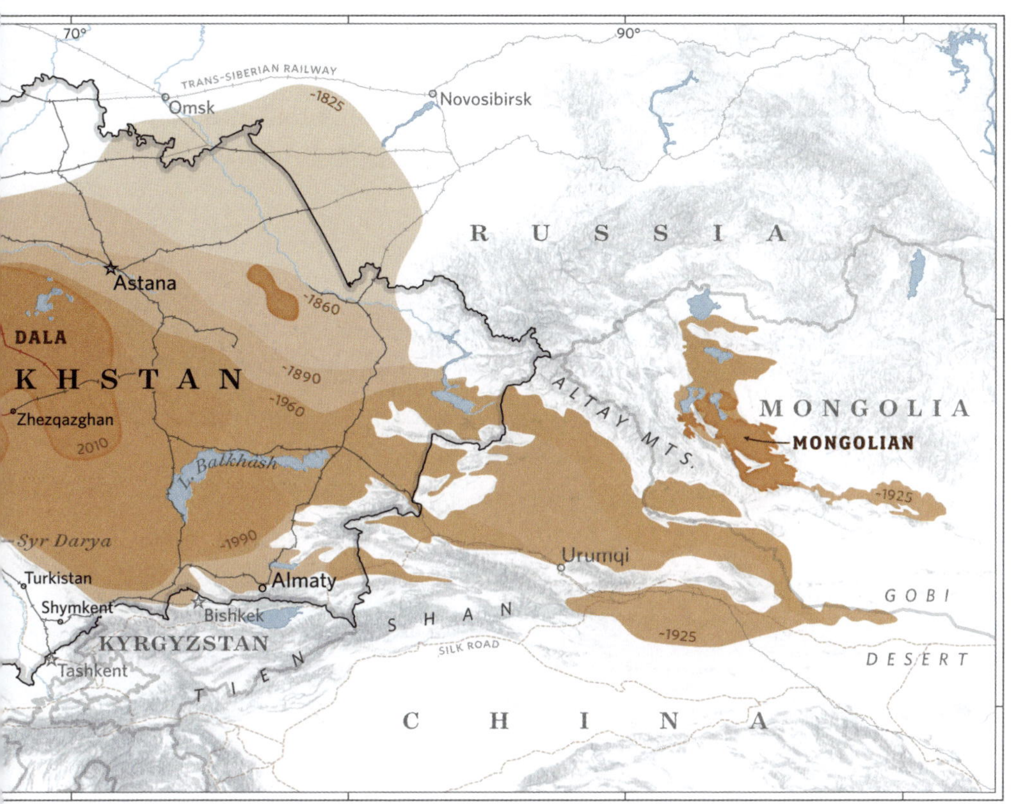

I said that indeed I found Kazakh history interesting, and had just visited Saraychik the day before, which was enough to prompt him to wave me into his room, where he pulled out a small antique silver coin. He poured black tea into a steel-rimmed mug and encouraged me to partake in a spread of bread rolls and candy he laid out on the table. Ah, I concluded, he was trying to sell me something.

"I don't have any money," I said through the app. When the phone's voice repeated my words in Russian, his face turned indignant and he shook his head. "This is not for sale," he said. "This is very valuable."

This man, whose name I learned was Viktor, was a hobbyist who enjoyed digging for ancient coins with a metal detector. He said he was surprised to see an American this far off the beaten path and asked what brought me to this part of Kazakhstan. I told him how I had just come from Kaztalovka, where I had been observing the local wildlife.

"We do not have wildlife in Kazakhstan," he replied. I clarified that I was visiting the saigas. "Ahhh, yes, this is good. Saiga are very important. They travel a lot, like me!"

Viktor had no roommates, so I decided to bring over some of the snacks from my room to add to his folding-table smorgasbord and spent much of the trip reading alongside him. I had brought with me a copy of *The Happy Isles of Oceania*, thinking that perhaps reading about Theroux kayaking through the South Pacific would provide the illusion of warmth on colder nights. Viktor, meanwhile, was reading not from a printed book, but from a huge stack of graph paper with handwritten script. He said he had transcribed this story many years earlier and took it with him on long journeys. He encouraged me to take pictures of his pages so that I could show them to a translator back home.

As the train meandered to the east, we passed the small town of Shalqar, where our path intersected with the new rail line that, as Albert's GPS data showed, severed the Ustyurt population's historical path to the Aral Sea. His team has been working tirelessly to unwind some of this damage by partnering with the Kazakh government to construct dozens of saiga-friendly crossing points, where slightly sloping dirt ramps rise up to match the height of the train tracks, making it appear from a distance as though there's a natural gap in the rail line. Ecosystem engineering that manipulates the landscape, not unlike the aran traps, but for more noble purposes. Cattle-guard fences positioned far away from the crossings ensure that trains and saigas never actually run into each other at these intersections, which would shatter the illusion of safety and inevitably cause the animals to abandon the area once again.

ACBK has also worked with the Uzbek government to reshape its border fences from a barbed-wire wall to an X-shaped barricade that blocks vehicles from driving through but leaves just enough space for saigas to pass. Over time, Albert hopes such crossings can rebuild the trust necessary for them to confidently return to a plateau they once relied on.

Early tracking data from other parts of Kazakhstan where similar reparations have been made suggests that once the railroad ramps are built, antelope may start to cross again. If so, the Ustyurt saiga population will, at some point soon, likely return to a remote corner of Central Asia that their instincts and faded memories tell them is worth the hassle to get to. For many reasons, including genetic diversity and reduced human disturbance, it probably is. But as our carriage churned southbound along the Syr Darya, a major river that feeds into the Aral Sea, I was reminded of the fragility of the saigas' situation.

Once the fourth-largest lake in the world, the Aral began its precipitous decline starting in the 1960s when the Soviet Union diverted much of the Syr Darya and other rivers as part of an ill-conceived agricultural irrigation project. In some years, not a drop of the river reached the sea, ultimately pushing the Aral to the brink of collapse. In subsequent decades, what had been an inland refuge for humans and nonhumans alike shrank to a tenth of its original size. By 2007, it was surrounded by so much exposed lake bed and desert that what was once a single body of water effectively split into four small ones, including the North Aral Sea and the South Aral Sea—the latest indication that this land, once an unbroken expanse of communities and corridors, has become increasingly disjointed.

Humans largely abandoned the region when the once-mighty lake turned to dust. In his authoritative travelogue, *Apples Are from*

Kazakhstan, Christopher Robbins wrote that the sea "had retreated in some places by as much as 100 miles from its original shoreline, stranding ocean-going trawlers among sand dunes, and revealing sunken hulks. Camels now roam where ships once sailed."

The government has initiated a handful of projects designed to revitalize and preserve what remains of the Aral, but these belated efforts are fighting an uphill battle. The summer of 2014, the same year that the new rail line across the Ustyurt Plateau abruptly cut off the saigas' southbound access, was the first in recorded history when the eastern basin of the South Aral Sea dried up completely—a sign of the region's severe ecological distress. If this trend continues, it sets the stage for a potentially agonizing visual that few if any humans will be there to witness. The moment when saigas finally walk back to a familiar haunt after a decade or so of interruption, to find that the world is not as they left it.

EVERY HOUR OR SO, as we pulled into a small town, hawkers would wander onto the train to show off their watches, rings, lace, and other goods for sale, before hopping out the back end as we pulled away. By evening, we passed through Baikonur, home to the famed Cosmodrome, from which most Russian space flights are launched, including the one that sent the saigas' ICARUS receiver to the International Space Station. It was from here that Yuri Gagarin left Earth in 1961 and became the first human in outer space, kick-starting a global race that would crescendo in the decade that followed.

In doing so, Baikonur also became the latest intrusion into the saigas' world. The Cosmodrome itself, still active today, falls right alongside the perimeter of the Betpak-Dala population's current range, meaning that even transit to the stars can disrupt migrations here on Earth. To mitigate such clashes, ACBK has taken to sharing

GPS tracking data with Cosmodrome officials so that they can try to schedule launches during periods when the saigas are not within earshot of the rocket boosters. Such are the challenges when humanity follows in the footprints of other animals' small steps and then attempts giant leaps.

From this point forward, my distance from the saigas would only grow. I was headed southeast, to places where earlier generations once roamed but the modern world says they cannot go. After traversing our shared evolutionary journey, this is where our trajectories diverge, at least for now.

As the faint glow of our carriage coursed past the whistle-stop ruins of Turkistan and Shymkent, where untold numbers of pilgrims once rested their heads in local caravanserai, my phone regained service, and the ping of my email inbox caught my attention. It was a message from Dr. Lee Ki-sup (aka "Kisup"), a contact I had been in touch with ahead of my final destination, the demilitarized zone separating North and South Korea. I was not scheduled to arrive in Korea for several days, but Dr. Lee's email said that if I could get there earlier than planned, it was possible to go to some parts of the DMZ that would otherwise be inaccessible. To participate, I would need to move up my itinerary and be in Paju, South Korea, in a little under forty-eight hours.

After taking a moment to process that new information, I committed to seizing this unforeseen opportunity and booked a ticket on the next flight to Seoul.

Early the next morning, we rattled into the Almaty train station and I hailed the first taxi to the airport. I had plenty of time to get there before my flight, but mentally, I was already gone.

At the terminal, my laptop regained internet service for the first time since leaving Atyrau, clear on the other side of the planet's largest landlocked country. With some time to kill before departure, I backed up my photos into a cloud server, in case things went sideways in the days ahead. Between shots of ancient artifacts and saigas racing over the frozen plains, I stumbled upon the pictures I had quickly taken of Viktor's graph paper and promptly uploaded them into an online translator. Within seconds, I was reading page after page of handwritten notes from someone I had only fleetingly crossed paths with, but whose manuscript felt hauntingly resonant of the saigas' tale.

His story, it turns out, is of an immortal man who elicits strong curiosity among mere mortals. They subject him to experiments to try to determine how he has acquired such a superpower. Every time he is released, he relocates to a new home where he can "return to a quiet place like a grandson of the night." To escape this cycle of constant disruption, the man briefly takes refuge in a deep cave. But this life of isolation does not suit him, and so, the narrator recounts, the immortal summons the strength to return to the surface.

> As time passed, you observed the changes in the world around you. The technology advanced, the cultures evolved, and yet, you remained the same, an immortal trapped in an ever-changing world. You became a silent witness to history, unable to age or die, but constantly adapting to new eras.
>
> You wondered if you would ever find peace, or if your existence would be an endless cycle of hiding and surviving. The burden of immortality weighed heavily on your soul, as you watched the world move forward without you. You longed for companionship, for someone who could understand your plight, but you feared revealing your secret to anyone.
>
> In the end, you realized that immortality was not a blessing, but a curse. You were condemned to an eternal

life of solitude, forever hiding from those who would seek to exploit your gift. You continued your journey, seeking a place where you could find solace and acceptance, but always knowing that your true nature would keep you apart from the world around you.

— 6 —

No Man's Land

Latitude

GAZING OVER THE wings as we descended into Incheon, I watched with half-open eyes as our mighty aircraft flared slightly backward, gliding ever deeper into a cushion of cool sea air. Gracefully tilting to and fro, our plane coasted for a few lovely minutes before carefully calibrating its final approach, extending its sturdy legs, and bracing for impact.

As I sprinted through the airport, I typed, "I need to get here as soon as possible" into Google Translate and copy-pasted the street address. After I showed it to the first taxi driver I saw, he grunted in the affirmative and we were off.

En route to our destination, I provided live updates to Kisup, who had warned me that they would depart for the DMZ at 11 a.m. sharp. The taxi GPS pegged my arrival at 11:16, so through my translation app, I asked the driver if he could possibly get there faster. Hearing the app's robotic Korean voice, he nodded firmly and hit the gas.

Screeching up to a generic four-story office building a few minutes after the top of the hour, I generously tipped the driver, grabbed my bag, and leapt out of the car. Kisup met me on the sidewalk and escorted me into the foyer of the DMZ Ecology Research Institute, where I was greeted by a group of very confused faces. A dozen people were seated around a long wooden table in the middle of a cramped room overflowing with dog-eared logbooks, potted plants, and landmine warning signs.

"I just have to clear the air on something," said an older gentleman in a lilting Canadian accent, puncturing the silence. "I didn't know you were coming here. Are you hoping to join us or what?"

This man's reputation preceded him. It was George Archibald, famed ornithologist and cofounder of the International Crane Foundation (ICF). Seated beside him and looking equally puzzled were Crawford Prentice, ICF's newly installed Asia director, and at least a half dozen Korean researchers and volunteers.

I said that I had been in touch with Kisup about observing cranes in the DMZ for a book I'm writing. I added that while I was not originally scheduled to arrive until later in the week, I was told that if I got here sooner, we could potentially visit additional sites. Standing nearby and absorbing the conversation, Kisup looked on sheepishly, apparently just realizing that he had been holding parallel conversations with two international visitors without fully briefing one about the other.

"You just arrived in Korea?" George asked. My plane landed an hour ago, I said, which spurred the Koreans in the room to do some quick mental math on how fast I must have driven to be standing in Paju sixty minutes later.

"I simply don't know how you can fit into this," George replied with polite exasperation. This was a very important trip, he explained, because he was trying to cram Crawford's head with as much knowledge as possible about key places cranes rely on within the world's most heavily fortified military border, before his organization made a leadership change.

"If you're willing, I would be very grateful to be a fly on the wall," I said. "I have no set itinerary and am happy to make my own arrangements for accommodations." Searching my mind for any other redeeming qualities I might offer, I conjured my best closing argument: "I have a backpack full of snacks, so I require no food."

Looking around the room for signs of objection, George found none, and the fog lifted on his wide grin. "I think it's wonderful that you're writing a book about this. We'll make it work."

Having settled that, we piled into a waiting SUV and hit the road. After a brief drive, we rolled up to a security gate lined with rifle-toting soldiers in army fatigues. These would be the only other humans we'd encounter the rest of the day.

CRANES HAVE LONG occupied a special place in East Asian culture and folklore, where they are seen as the living embodiment of wisdom, prosperity, good fortune, and longevity. The innumerable rice paddy fields, mudflats, and coastal marshes that once dotted the Korean peninsula would overflow every winter with slender and noble birds, whose annual migration ecologist Aldo Leopold described as the "ticking of the geological clock."

Given the choice, the red-crowned crane, conspicuous and statuesque at five feet tall, with a distinctive patch of crimson skin popping from an otherwise black-and-white frame, would gravitate toward the secluded valleys of the hilly north. The slightly smaller white-naped crane, with an ashen body and salmon legs, preferred the wide-open spaces of the southern lowlands. For most of the past two thousand years, both landscapes were available in great abundance.

Tight-knit crane families, generally composed of a mom, a dad, and up to two precocious children, could be seen strutting around all day as part of their rigorous foraging ritual. In the evenings, the families would converge on large overnight gathering sites, which proffered the literal and figurative warmth of community, as well as increased security and copious chatter about the important matters of the day. In nearby fields, flocks of amorous adolescents and single adults provided a sizable dating pool from which bachelors and bachelorettes could find their lifelong partner. Their lives, fluid and serene, had balance.

Then, in March 1950, as the weather began to warm, the snowbirds departed for their summer homes in northeastern Asia. While they were away, the world changed.

That summer, North Korean soldiers surged across the 38th parallel, which had stood as a temporary border since the end of World War II, keeping the peace until a unified Korean government could be established. Not only did unification never come to pass, but over the next three years the peninsula erupted into a burning hellscape that killed more than 2 million people, displaced millions more, and caused catastrophic damage to the region's infrastructure. By the time the hostilities ended with a ceasefire armistice, both sides had been thoroughly obliterated. A formal peace treaty was never signed—only an agreement to pause attacks—which means, strictly speaking, the two sides remain at war.

In the wake of this grueling campaign, a 160-mile-long demilitarized zone was carved through the middle of the peninsula, as a barrier to prevent further clashes. Cutting through the DMZ's center is the Military Demarcation Line (MDL), which is the actual divider between the two countries. A little over a mile on either side stand the DMZ's northern and southern boundaries, behind which each country's army must remain.

Just below the southern boundary, South Korea also established an additional buffer area called the Civilian Control Zone (CCZ). Like the rest of the DMZ, the CCZ is strictly off-limits to the average citizen, though it can sometimes be accessed by nonmilitary personnel with government permission. To restrict access, the CCZ is walled off by military checkpoints, observation towers, and fortifications. (North Korea operates no CCZ equivalent on its side.)

In the aftermath of the war, communities on both sides faced severe food shortages, psychological trauma, property damage, and displacement. Many families were cleaved apart, stranded on opposing sides of the DMZ, and whole cities had to be rebuilt from the ground up.

During the three years of battle, many of the wetland homes the cranes once frequented were blown to bits or subject to active fire. George believes the birds likely sought refuge on the abandoned farms of soldiers who left for the front lines.

By autumn of 1953, the bombs had stopped falling, but in the subsequent decades after soldiers returned home, the human population on the Korean peninsula more than doubled, from around 30 million people to more than 78 million. In the South, this rise was accompanied by extraordinary economic prosperity that rapidly replaced marshes and rice fields with high-rise condos, freeways, electronics factories, office complexes, shopping malls, and golf courses. Places about as suitable for cranes as the moon.

In the communist North, economic growth has been much more anemic. Periods of deep famine have also devastated local wildlife populations as desperate people sought to stave off starvation by snaring and gunning down every living creature they could find. During the so-called Arduous March that followed the collapse of the Soviet Union and the global network it supported, it was not uncommon for North Koreans to survive on corn husks, grass, or tree bark, which was stripped and boiled to make porridge. With the land so thoroughly scraped of natural resources, animals who did manage to avoid capture often starved anyway.

The one place on the peninsula that did regenerate was the DMZ, which was packed like a powder keg with more than a million land mines, and where human inhabitation and development was strictly limited. Emerging from the rubble, trepidatious wild boars, mountain goats, Asiatic black bears, and musk deer began to reappear after years caught in the cross fire. Locked in by high-voltage electric fencing, the mammals' homeland had in short order become an open-air prison, where they could live unencumbered but never leave.

For the birds, who float effortlessly above the barricades and are generally too lightweight to set off unexploded ordnance, the DMZ became a haven of last resort. A narrow sliver of land that, for reasons wholly unclear to them, had been left untouched. Communities that once roamed their own private corners of the peninsula soon converged, seeking asylum in one of the only remaining spaces available to them and forging something akin to a United Nations of avian societies.

In this pulsating microcosm of Korea's past, Oriental storks, Mandarin ducks, white-fronted geese, and black-faced spoonbills, each with their own distinct history and communication style, share space alongside cinereous vultures and Steller's sea eagles, whose eight-foot wingspan casts ominous shadows over the forsaken land. Among the hovering carnivores, periodic land mine detonations by deer and wild boars only make the DMZ all the more appealing. Some border guards swear they have seen eagles intentionally drop their prey onto an explosive to make the flesh easier to pick apart.

As Korea's traditional wetlands were drained and repurposed, this accidental sanctuary became critical to the survival of many birds, but none more so than the majestic red-crowned and white-naped cranes—once distant acquaintances, now neighbors of necessity. Of the four thousand red-crowned cranes left on Earth, more than half spend time in the DMZ. Among the planet's twelve thousand white-naped cranes, 90 percent spend at least part of the winter surrounded by machine guns and barbed wire.

THE NEED FOR one's own space, whether along the 38th parallel or elsewhere, is fundamental to every animal society. It affords individuals and groups the freedom of action, or *latitude*, to conduct life as they see fit. But how much space and what that space must include or lack varies enormously. Some animals, like drywood termites, can live their whole lives inside a single log, but require large population numbers and complex social structures to survive. Others, like gray wolf packs, travel in small groups, but can require vast home ranges that span hundreds of square miles across, based on local prey availability. Given adequate space and resources, animal nations can rapidly swell in size, shaped by disparate social groups loosely linked through network connections and shared histories.

Conversely, the destruction of spaces other species rely on can have ripple effects throughout the animal kingdom. Agricultural land-use change is by far the biggest modern-day driver of wildlife displacement, and research shows that humans have already transformed or occupied roughly 75 percent of the world's land, leaving less room for others with each passing year. As the amount of livable space dwindles, human and nonhuman populations become increasingly compressed, creating opportunities for conflict and competition over finite resources within and between species.

Displacement also occurs in areas where humans have little or no footprint. If an elephant knocks down a tree to gain easier access to its tasty leaves and roots, it helpfully disperses seeds and contributes to greater biodiversity. But the arboreal animals who previously inhabited the fallen tree's canopy will have to branch out and find a new place to live.

Sometimes, places are radically altered by factors beyond anyone's control. Roughly three million years ago, when the volcanic Isthmus of Panama rose up from the seafloor, it created a new land bridge between North and South America and simultaneously closed a transit corridor between the Caribbean and the Pacific used by countless aquatic animals. This seismic shift ushered in a grand redistribution known as the Great American Biotic Interchange, which would forever alter the global ecological balance. Suddenly, flightless birds started encountering wolves, bears, and other predators they'd never had to face before. South American native ungulates were forced to compete for resources with newer arrivals who descended upon them from the north.

There's an element of unpredictability in such disruptions. When a hurricane ravages an area, it could drive animals to fundamentally restructure their societies, like the rhesus macaques on Cayo Santiago, or completely abandon a place, as Merlin Tuttle saw with the lost cities in Florida bat caves. Sometimes, birds and insects will get trapped in the eye of a storm and choose to move with it—a phenomenon radar-watching meteorologists call "bioscatter"—before making a Dorothy Gale exit in some distant and unfamiliar destination. One needn't look any further than the warring factions of Argentine ants to see how that can play out.

No matter the circumstances, the continuous push and pull of animal populations—human or otherwise—into new spaces creates opportunities for nations to rise and fall.

AFTER CLOSELY EXAMINING our personal documents and reconfirming the intentions of our visit with Kisup, the border guards cleared us to enter the CCZ just north of Paju. As we slowly rolled over the Unification Bridge spanning the ice-cold Imjin River, our passports stayed behind on the south bank, held tightly by the guards until our promised return.

Crossing into the CCZ is a time-warping experience akin to entering a museum diorama. It transports visitors to an alternative version of Korea's agrarian past in which all the farmers mysteriously disappeared the day before. An ambient tension hums over tidy rows of exhausted rice crops and flapping tarpaulins.

The plots are tended by fieldworkers who live outside the CCZ but are allowed to periodically visit thanks to special government permits. Some families make additional income from growing ginseng or other crops, and greenhouses increasingly dot the landscape, much to George's dismay. Others lease their land to construction companies that need a cheap place to dump excavated dirt. Officially there's a height limit of 50 centimeters (20 inches) on how much dirt can be dropped, but the authorities have no motivation to enforce such arbitrary rules.

Besides, the real value is in the land itself. CCZ landownership information is not public, but a significant portion is believed to be owned by the government, which then leases it to farmers. The rest is owned by private individuals, who pass it down as a nest egg through the generations, clinging to faint hopes that someday their dusty parcel will form the foundation of a glorious new skyscraper in a unified Korea. The sunken cost of waiting decades for this anticipated windfall has been enough to spur the deed owners to stymie every attempt to turn their land into a permanent wildlife preserve.

As we crawled along a grid of single-lane roads lined with red-

and-yellow land-mine warning signs, Kisup kept the car windows down and listened to the breeze. Since 2002, he has spent every winter conducting block-by-block crane censuses in the CCZ, adding small drops of color to parts of the map that would otherwise appear completely blank. It's yeoman's work that he carries out with a stoicism and precision bordering on obsessive compulsion. After completing each survey block, he cross-checks his data. If there's even a slight chance that some of the cranes got mistakenly double-counted—a distinct possibility given that birds are constantly flapping from place to place—he will go back and count again. "You're very thorough," I remarked, extracting only a silent nod in response.

Examining a series of pinpoints from his previous surveys, he glanced upward, grabbed his binoculars, and encouraged me to do the same. In a distant field, I then got my first sight of a secluded family of red-crowned cranes. The father was out front, standing nearly as tall as I am, with his partner just behind, followed by their two children. Considered the most iconic of all birds among Koreans and Japanese alike, red-crowns are often referred to simply as "the crane"—with all other cranes relegated to second-class status. Superficial as that distinction may be, it was easy to see how one could be captivated by the sight of such strikingly regal figures.

Mirrored in the field next door was a white-naped crane family of equal size. While cranes are highly territorial within their own kind, frequently shooing competitors away from desirable food sources, the adjacent white-naped and red-crowned families complement each other, George said. They have slightly different dietary preferences, with the white-naped preferring seeds, roots, and grains, and the red-crowned opting for a more omnivorous diet of insects and other small animals. When a predator is on the attack, the birds will often join forces to fight back.

As he alternated between coasting and looking through his binoculars, Kisup took care not to hit the brakes. To the birds, a slow-rolling passerby is a reason for alertness, but a vehicle that comes to a full stop is a predator watching its prey.

The cranes are keenly aware of where the CCZ starts and ends, Kisup said, and are careful to stay within its boundaries—not that they have much choice. In a country where undeveloped land has dwindled in recent decades, the lots to the south and west of the DMZ, once deemed undesirable for their perilous proximity to enemy territory, have been steadily compressed by the encircling walls of modernity. Paju, the region the cranes look at just across the river, has tripled in population density since 2000, from 178,000 people to more than 600,000. To the west, the glittering city of Gimpo, located near the biologically rich confluence of the Han and Imjin Rivers, was a small town until 1998, but is now home to more than 800,000 people.

Cruising farther down the road, we passed many more crane families, each of which was painstakingly logged in Kisup's records as he counted out their numbers and identifications. "Three white-naped. Four red-crowned. Eight red-crowned. Forty red-crowned. Seven white-naped . . ."

When he first started doing this work more than two decades ago, the lanes inside the CCZ were not paved, and off-road travel remains suicidal. Since drones and other technologies that would normally help researchers access places farther afield were a nonstarter in a military zone, Kisup instead cultivated relationships that granted him access to a number of elevated guard posts and prime overlooks from which he could stare down into terra incognita.

One of his more successful lookouts is a small hilltop cemetery, which was created for North Korean defectors who had fled to the South but requested to be buried close to home. Stepping out of the car next to a windowless, robin's-egg-blue mausoleum, we walked among a dozen or so engraved slate tombstones, some of which were relocated here from ancient grave sites to make room for new mili-

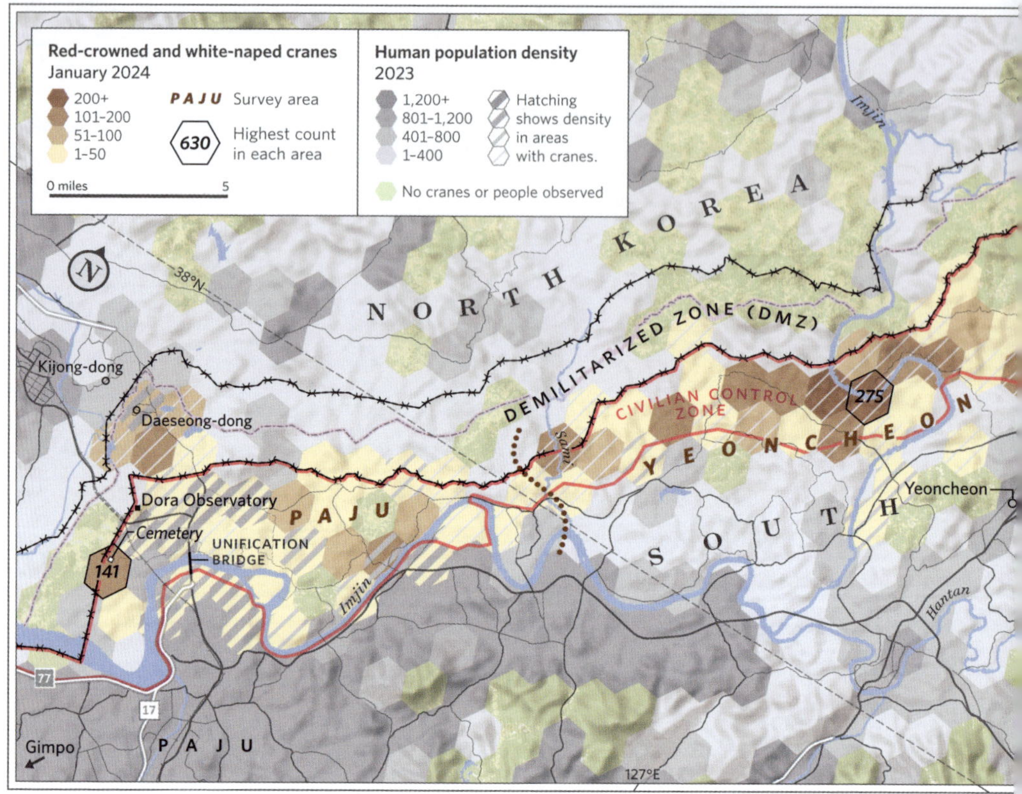

tary infrastructure. Taking in the panoramic view of rippling, dried-out hills and riverbeds that ran uninterrupted to the edge of the CCZ, Kisup could see crane families far off the beaten path. "Three white-naped. Fifteen red-crowned. Six red-crowned . . ."

The nearby Dora Observatory, perched high on a hilltop south of the 38th parallel, has also been a primary post from which he spots cranes living off the grid. Taking in the view from its summit, I was struck by how remarkably narrow this strip of undeveloped land really is. From a single spot, visitors can see both the South Korean village of Daeseong-dong and the North Korean Potemkin village of Kijong-dong, barely a mile apart. The two villages were permitted to remain in place as part of the 1953 armistice and are perhaps best known for their juvenile one-upmanship. In the '80s, when the

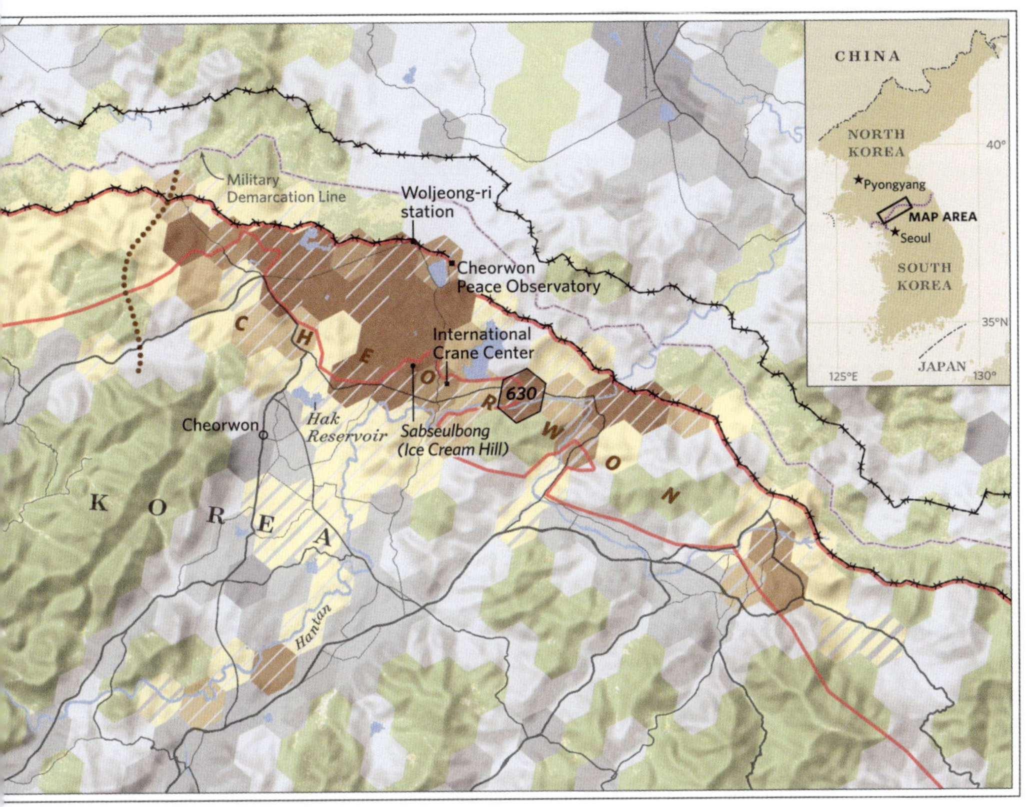

South built an enormous flagpole more than 300 feet tall, the North responded in kind by constructing one that was 200 feet taller. To match the gargantuan pole, the flag on the northern side weighs nearly 600 pounds, so heavy that it hangs limp most of the time—surely a metaphor for something.

Far above this petty skirmish, two cranes soared effortlessly above the divide.

From this high vantage point, George tuned his ears to the breeze and eavesdropped on the chatter down on the ground. Birds and many other animals use all manner of vocalizations to communicate with one another. In their book *Animalkind*, Ingrid Newkirk and Gene Stone document how "even the most arbitrary-seeming squawk has far more nuance and purpose than we originally thought." The American robin, for instance, may use a variety of different calls to warn his flock of potential threats. A "sharp 'tic tic'" call could indicate a prowling cat nearby, but a "thin 'seeee'" means he's spied a hawk. "Until recently, scientists believed this 'talking at' rather than 'talking with' model was the full extent of animal communication," but now they're not so sure, the authors wrote. Citing a peer-reviewed 2018 study published in the prestigious journal *Philosophical Transactions of the Royal Society B: Biological Sciences*, Newkirk and Stone suggest that some animal sounds we hear may, in fact, be two-way conversations—a trait once mistakenly assumed to be uniquely human. Among humankind, they note, there is typically a two-hundred-millisecond gap between turns in a conversation, because that's the time it takes for the brain to recognize the other person has stopped talking, prepare a thought, and respond. By contrast, songbirds reportedly wait roughly fifty milliseconds to reply during a conversation, "while the more measured sperm whale collects his thoughts for two full seconds before responding."

Crane societies have their own well-developed series of kazoo-like calls, George said, deployed for specific occasions. And as the world's leading authority on cranes, he has become something of a foreign-language interpreter who can personally recognize at least sixteen different calls. These include a precautionary warning call; a more urgent alarm call; a preflight call; a flight call; a territorial threat call; a lonesome "location call" that reunites separated contacts; a nest-building call; a contact call used when birds are standing side by side ("kind of a low *brrrr* sound"); a pre-copulatory call ("more of a scream, really"); and a unison call, which is actually a duet in which a male and female produce complementary harmonies to deter other birds from their area and signal to their mate that it's time to take their relationship to the next level.

Like many animals, cranes are also believed to have distinct regional dialects and unique family calls, George added. This means that with a well-trained ear, one can learn quite a lot about the history and status of a particular bird community simply by listening to the airwaves.

SCIENTIFIC MONITORING OF large birds is among the most advanced segments of the animal-tracking space, in no small part because it often leads to unexpected results.

In the early 1800s, a German count named Christian Ludwig Reichsgraf von Bothmer shot a stork while on a hunting trip. Upon retrieving the bird, he was stunned to find that the animal had an entire wooden spear protruding vertically out of his or her body. The count took the bird to a local professor, who identified the spear as being African in origin, signaling that the stork had traveled a remarkably long distance before being blasted out of the sky.

This assessment struck the count as highly unusual, since many leading theorists of the day believed that birds lived in a fixed area and hibernated in the wintertime, possibly underwater, which explained why we didn't see them. Zoologist Lucy Cooke notes that competing theories of this era posited that birds underwent a winter transmutation, morphing into mice or plants during the colder months before returning to their familiar birdlike form in the springtime. The idea that such fragile creatures would have the stamina to travel en masse across vast bodies of water—after being skewered by a spear, no less—seemed completely absurd.

Yet in the years that followed, the evidence for transcontinental migration kept mounting. Dozens more birds were found throughout northern Europe with African spears jutting out of their bodies, so many that they were given a name: *pfeilstörche*, or arrow storks.

Intrigued by the prospect of birds making such vast odysseys, a fellow German named Johannes Thienemann joined forces with an army of local volunteers to affix aluminum rings around the legs of thousands of storks, each of which included a unique ID number and his office's mailing address. In the subsequent years, more than 170 of the rings were individually returned to him through the postal service, a quarter of which came from Africa. Informed by the data he received, Thienemann was able to plot the migratory patterns of storks with an unprecedented level of precision.

Leg-banding is still a common practice in scientific circles, Kisup said, which is how he knows that many of the individual cranes he logs have been coming to the CCZ for as long as he has. He's confident they remember him.

Technological advancements in geolocation and satellite tracking have also added new dimensions and textures to our avian observations. It was through such methods we learned that tiny arctic terns, weighing all of four ounces, not only zigzag more than 44,000 miles between Greenland and Antarctica each year, but that they do so by riding specific air currents that ensure they will never need to fly into the wind.

The longest-running GPS crane tracker on record is affixed to Borzya, a white-nape named after the river where she was tagged by Mongolian and Russian researchers back in 2015. That first winter, Borzya migrated south to Poyang Lake, traditionally a major destination on the southern side of China's Yangtze River and the end point of what's known as the western flyway. Amid China's economic development, this flyway now traverses some of the most heavily populated and industrialized landscapes on Earth and birds face many challenges along the way, including power line collisions, habitat loss, and poisoning.

Air pollution is also a major problem for birds flying over industrial zones around the world. When two University of Chicago researchers assessed more than one thousand birds who flew over

America's Rust Belt during a 135-year period, they observed that during years of increased fossil fuel usage, such as the manufacturing boom during World War II, the feathers and bellies of white and yellow birds were stained black from flying through dark particulate matter. Such extra-fine particles can burrow deep into the birds' lungs, weakening their immune system.

So perhaps it should not be surprising that Borzya didn't return to Poyang Lake the following year. Instead, she changed course, shifting to the eastern flyway, which skirts down through the Korean peninsula toward southern Japan. To date, Borzya is the only white-naped crane to have been shown to make such a flyway pivot, but experts speculate that it happens much more frequently than we know.

When satellite data showed that Borzya arrived in a restricted area of Paju, Mongolian researchers knew they had to call Kisup, who promptly went out to visit her. He watched as Borzya flitted around between the DMZ and the surrounding wetlands, where she joined up with hundreds of other white-naped cranes. She has been back every year since then, Kisup said, though he had not yet received a status update from Mongolia this winter. He hoped she was on her way.*

As we wove back through the rolling hills toward the CCZ entrance gate to collect our passports, George was transfixed by the sun descending behind the distant mountains. Another day had come and gone, and across the DMZ, the cranes were returning to their evening roosts to tuck in for the night. Having barely slept since stepping off a Kazakh train, I was eager to do the same.

The others in my group had dinner plans with a coalition of

* One month after our visit, Kisup finally received a biologging update from Mongolia, which showed that indeed Borzya had arrived in a Paju hotspot just east of the Dora Observatory a little over a week before we did. It's entirely possible we were looking right at her and didn't know it.

Korean conservation groups, which they invited me to join. I politely declined, opting instead to melt away in the most tranquilizing bath of my life.

THE NEXT MORNING, I met up with George and Crawford for breakfast in the hotel restaurant, which offered a panoramic view of the Imjin River and the mountains of North Korea in the distance. We were the only people in the restaurant and quite possibly the hotel's only guests, which may explain why Michael Bublé's Christmas album was serenading us through the loudspeakers.

Peering out over the river, George asked if I noticed anything unusual about it. I scanned across, looking for signs of flooding or other natural deviations, not finding any.

"It has no boats on it," he remarked—a highly unusual sight given its proximity to major South Korean cities. "Should peace come, this would be one of the busiest rivers in the world. Twenty-two million people live right there"—he pointed in the general direction of Seoul—"but if any of them took a boat out on that water, they'd be shot dead." This dark reality has created favorable conditions for the cranes, he added, who would no doubt be driven away if the river suddenly filled with grumbling cargo ships and belching barges.

By midmorning, Kisup arrived in the lobby to pick us up. Since each section of the CCZ has a different level of security, and entrance approvals are granted on a case-by-case basis, one cannot simply drive horizontally from one area to the next. As we moved farther east, we had to exit and reenter every time. This morning, we were headed north of Yeoncheon to a zone seldom visited by civilians.

Due to government sensitivities, security standards in Yeoncheon are considerably stricter than in Paju. At the checkpoint, we were provided military escorts—or minders, depending on who you ask—who followed our every move. Photography of any kind is strictly prohibited, even for researchers, so guards put stickers over our phone lenses and hung a thick burlap bag on our rearview mirror so that we couldn't access our vehicle's backup camera. A blue flag affixed to the top of the driver's side window signaled to enemy watchtowers that we were a neutral party.

Having been decked out in the necessary don't-shoot-us apparel, we rolled slowly down the road, which was lined with high-voltage fencing and massive coils of barbed wire. The triangular land-mine warnings were identical to the ones we'd seen in Paju, but somehow looked more menacing in this setting, as though they really meant it this time. Spotting a water deer on the shoulder ahead, we paused and watched as he took precise and deliberate steps into the woods. "He knows," Kisup said.

After clearing a series of additional security checks and watchtowers, we parked on the edge of an overlook that stared down at a lush riverbank, where huge clusters of cranes were congregating. Peering through his powerful binoculars, Kisup murmured as he counted to himself. Fifteen seconds later, he pulled his hands down and said, "One hundred and thirty white-naped, fifty red-crowned."

Explaining his method, Kisup said he had perfected the art of hand counting individual birds until he gets to ten, and then eyeballing how many ten-sized groupings he sees. For especially large groups, he will instead hand count one hundred birds and then ballpark how many hundreds there are. All of these eyewitness observations are logged, so that he can compare numbers from month to month and year to year. If only the soldiers stationed in the DMZ could be trained to count cranes, his job would be a lot easier.

As I walked back toward our car, I heard the unmistakable rustling and crunch of footsteps in the nearby woods—disquieting sounds to hear in a war zone. We turned to Kisup for guidance, chuckling nervously, mere moments before two wild boars bolted from the brush and sprinted manically up the road.

As we continued our journey, we rambled through the valley's terraced rice paddies, and Kisup resumed his counting spree. Three red-crowned and two white-naped here. One hundred and forty red-crowned and eleven white-naped there. As we spiraled upward, a small cluster of featureless buildings emerged on the hillside. Our minders instructed us to park our vehicle there, but said if we wanted to, we could go just beyond the complex on foot.

Walking alongside a large, blue-roofed outpost, we were reminded by the soldiers—not that we needed any reminding—that photography was strictly prohibited in this area. They then proceeded to nudge us down a sloping footpath. Down the hill, I could see that the path terminated a short distance ahead at an imposing steel barricade. Beyond that wall lay a vast, unspoiled valley. Sweeping in scale and devoid of even the slightest human imprint, the primordial land was bisected by the Sami, a majestic tributary flowing untamed through its floor.

"What *is* this place?" I asked, taken aback by the abrupt change in scenery.

That fence divides the CCZ from the military DMZ, Kisup explained. For a civilian to step on the other side of this barrier would be tantamount to an act of war.

At the bend in the river, a mile or so up, stood the Military Demarcation Line, identified by a series of yellowed posts. Beyond them lay an undeveloped slice of North Korea, whose military watchtowers were plainly visible to the naked eye on a near-distant hill. Handing me his binoculars, Kisup guided my eyes downward to a frozen sandbar on the river's west bank, equidistant between where I stood and the MDL. There, huddled along the waterfront, stood two hundred cranes, splashing around, blissfully unaware of the profoundly hostile space they have chosen to inhabit.

In the prewar days, there was a village at this riverbend, Kisup said, before it was destroyed and abandoned. Today, any trace of human settlement has been washed away, and the land has healed under the stewardship of its new feathered residents. When I listen back to the ambient audio recording my phone captured from this overlook, I hear only two sounds: the rustling wind through the trees and the calls of the cranes, which echoed through the canyon.

Down on the western bank, many of the cranes appeared to be dancing. As with their vocalizations, George explained, cranes have a wide range of moves they use to express themselves in specific situations, from greeting a long-lost friend to deterring unwelcome guests.

Most dancing begins, he said, by the bowing of heads, as an acknowledgment of their peers. If a red-crowned crane wants some-

one to back off, they can also engage what's known as the "bow threat," which combines a ritualized bow with a vertical flaring of their wings, creating a sharp contrast between the black-and-white feather wall and the stark red of their cap.

Conversely, when they want to show affection to their partners, they will softly coo and fully extend their wings, as if to visualize how much they love each other. Leaping, spiraling, head bobbing, feather ruffling, charging, tossing small objects in the air—every action has a clear intention and meaning.

Watching this Whoville scene from our hillside lookout, even George appeared overtaken by its vibrancy. He racked his brain for the right words to describe the sight, finding only "Beautiful valley. No people. Happy cranes." After pausing, he added: "There are many places that have been changed by the hand of man. But there are very few that were heavily developed for thousands of years and then suddenly everybody leaves."

Such circumstances do have some precedent. During the Cold War, the border area between East and West Germany became a de facto nature reserve popular among freedom-seeking otters, frogs, black storks, and white-tailed eagles—a legacy conservationists sought to continue after the Iron Curtain fell by rebranding the former "Death Strip" as the "Green Belt." The Amur River border between China and Russia remains remarkably well protected from encroachment, even as geopolitical tensions have subsided. South American jungles once off-limits due to threats by antigovernment guerrillas have been found to contain extraordinary biodiversity upon reopening to visitors. In a world replete with conquerors and claim-jumpers, nonhuman nations often gravitate toward the quiet in-betweens.

Wildlife has similarly flourished in the Chernobyl Exclusion Zone following the nuclear disaster that forced the evacuations of more than two hundred thousand people, creating space where there once was none. In our absence, the wolf population rose by more than 700 percent, due to reduced hunting pressures. Brown bears, two hundred species of birds, and elusive animals like the Eurasian lynx also resettled there in our wake.

Once animals establish temporary or permanent residency in a particular spot, that location often carries a new level of significance for future generations. Among whooping cranes—the taller North American cousin to the white-naped and red-crowned—sons often inherit a slice of their father's territory in the cranes' summer breeding grounds to maintain the genetic line and keep the land in the family. Tracking data also shows that many young males will return to the same place every winter to be near their parents. (Daughters disperse more widely, which prevents inbreeding.)

We don't know definitively whether other cranes follow similar practices, but George presumes they do. If so, looking down on that remote valley, we were witnessing new family traditions being created in real time.

THE NEXT MORNING, I was awakened in my hotel room by a phone call. I was in downtown Cheorwon, a town at the DMZ's geographical center near the easternmost stop on our journey. Kisup was on the other end of the line, speaking more quickly than usual.

He would not be able to pick me up this morning, he lamented, because his father-in-law had unexpectedly passed away the night before. He was needed at home, so he instructed one of his local colleagues to pick me up instead and take me to today's destination. I said I understood completely, and that was that.

After I hung up, it occurred to me that today was, in fact, the day that I was originally scheduled to meet Kisup for the first time. Had I not scurried off that Kazakh train, my itinerary would have had me traveling north from Seoul to Cheorwon to get my first glimpse of cranes in the DMZ—plans that would have almost certainly unraveled under these tragic circumstances.

Feeling grateful to have connected with Kisup when I did and with a little over an hour to wait before his associate arrived, I strolled the streets around my hotel, where I was struck by an odd trend among the storefronts. I passed by a coffee shop, a family-photo studio, and then an office space for picking up prepared meals. In all three instances, the businesses advertised that they were open twenty-four hours a day, but curiously, none had any employees. These stores were fully automated, part of an escalating trend of robo-businesses that rely on customers to place their orders—whether for a soy latte or a Wild West photo shoot—through an iPad or tablet, while closed-circuit TV cameras watch from overhead.

Such businesses became increasingly common in South Korea in the wake of a pandemic-era labor shortage, but the outbreak only accelerated an inevitable outcome in a country that—following decades of uninterrupted, postwar population growth—now has one of the lowest birthrates in the world. Coincidentally, while sitting on one of the robo-café's couches and exploiting its free Wi-Fi, I was reading *The New York Times* on my phone, which that morning published an op-ed on this very topic. In it, columnist Ross Douthat put the implications of South Korea's plummeting birthrate—which in 2023 fell to around 0.7 children per woman—in context:

> A country that sustained a birthrate at that level would have, for every 200 people in one generation, 70 people in the next one, a depopulation exceeding what the Black Death delivered to Europe in the 14th century. Run the experiment through a second generational turnover, and your original 200-person population falls below 25. Run it again, and you're nearing the kind of population crash

caused by the fictional superflu in Stephen King's "The Stand."

In some ways, the shrinking of the human population would appear to be welcome news for cranes and other animal neighbors. After all, it was the postwar baby boom that caused so many wetlands to be drained in the first place to make room for glistening new condos and golf courses. But robo-businesses like these test that theory, because even in places with no consistent human presence, our impact persists. Unlike the war-torn villages of yesteryear that were swept away over time and returned to a blank slate, these zombie storefronts are unoccupied spaces that nonetheless prevent others from moving in.

RIDING ALONG WITH Kisup's colleague, Park Jin-young, she and I pulled into the International Crane Center, a sleek new museum where ICF also maintains a local office. Situated just outside the CCZ, the center used to be an elementary school, an elderly local told me, before it was shuttered because "the young people are not giving us children anymore." Inside the museum, there are educational exhibits, a library, screening rooms, and ticket booths for guided birdwatching tours. Next door, ICF owns a small dormitory for visiting guests, which is where I found George and Crawford.

George, a dutiful Christian, had just finished his morning rituals. He begins every day by studying Bible lessons, which can sometimes take up to two hours. Afterward, he listens to some of his favorite inspirational songs. He also prays for "little miracles" to occur throughout the upcoming day, and reflects on the heavier topics on his mind. That morning, his prayers were focused singularly on Kisup's family. When I asked him whether he sees any connections between his faith and his crane advocacy, he simply replied, "Every connection."

With Kisup unexpectedly out of commission, George suggested that we adjust our plans for the day and take a stroll along the boardwalk around the nearby Hak Reservoir. *Hak* means "crane" in Korean, but the water is popular with all kinds of birds. Spotting a group of swans clustered on the far side of the frozen pond, George live-translated their body language: "Hello! What a long flight." "So good to see you again!"

The Cheorwon Basin is an epicenter of bird activity in Korea, in large part due to its topography. At eight miles across, it's the largest open area in the central highlands and features several spacious freshwater lakes. Many of its current avian residents previously wintered in an enormous wetland area near the airport, until construction of a new highway across it began in 1991.

Road noise is one of the most persistent deterrents for cranes and other birds because it smothers the natural sounds the animals rely on to safely navigate their world. Researchers in Idaho once demonstrated this by setting up a half-mile corridor of speakers and playing looped recordings of passing cars—a "phantom road"—in an area renowned for bird migrations. Even without the visible blight of passing vehicles or stench of tailpipe exhaust, the disembodied sound was sufficient to drive a third of the birds away, science writer Ed Yong reported in *The Atlantic*. Many of those who stayed paid a price for persisting. "With tires and horns drowning out the sounds of predators, the birds spent more time looking for danger and less time looking for food. They put on less weight and were weaker during their arduous migrations."

That sensitivity to disturbance is what makes Cheorwon a precarious paradise, George said. Should reunification happen between North and South Korea, this basin would be among the first places to capitalize on it. All around town, signage describes a "land dreaming of unification," with artistic renderings of new soccer fields, office parks, manicured gardens, and riverfront condos. In the center of the plateau stands the bombed-out Woljeong-ri train station, including a mothballed ticket office, short segment of rusted tracks, and the charred-metal remains of an ex-train. This station once connected Seoul with the port city of Wonsan, now located in North Korea, and locals openly fantasize about its eventual return to glory. At the abrupt end of the Cheorwon line, a sign reads THE IRON HORSE WANTS TO RUN AGAIN.

As recently as 2019, there appeared to be a very real chance of that happening. When President Donald Trump met with North Korea's Kim Jong-un for a diplomatic summit in Vietnam, among the outcomes on the table was an easing of US sanctions against the North in exchange for tangible steps toward denuclearization. Among the first projects set to be funded if that deal was signed, according to South Korean president Moon Jae-in, was the reconnection of inter-Korean rail links. In time, this could allow South Korea, which has effectively been a railroad island since the '50s, to once again connect to China by land and transport goods throughout the Asian continent.

The deal ultimately fell apart, due to what the US deemed unreasonable demands from the North Korean side, and in 2022 President Moon was replaced by a more hard-line South Korean president who squashed any plans for a deal with the North. Woljeong-ri station would, for now, sit in wait.

Among the landowners who have held on to small rice plots within the Cheorwon portion of the CCZ, it's all part of a familiar push and pull. A seemingly endless churn of shifting political tides. But many still cling to the hope that someday their family will be able to cash out as part of a grand regional transformation, even if they don't live long enough to see it with their own eyes.

George, meanwhile, hasn't given up hope that the landowners

can be persuaded to keep the plots exactly as they are. Kisup's team has worked with the Cheorwon government to provide compensation to local farmers in exchange for the landowners leaving their leftover rice straw out for the cranes to eat, rather than selling it as cattle feed. George also instructed Crawford to commission a land-use study to learn precisely which families own what, so that conservationists can persuade them to sign over their land as part of a designated preserve. He knows it's an uphill battle, because the value the landowners would get from turning their plots into a protected bird sanctuary will never come close to what they could earn from a condo developer. But the checks for a nature preserve could be cashed now, compared to having a theoretical payday in the distant future. For some, that might be enough.

George asked where I was planning to stay that evening. I hadn't booked anything, I said, though perhaps I could go back to the downtown hotel Kisup had arranged for me the night before. Nonsense, he replied. They had plenty of beds at the ICF dormitory, which I should feel welcome to use.

He also shared that they had plans to go to some very special places the next morning and offered to make arrangements so that I could tag along, which I promptly accepted. "You won't regret it," he said.

RISING IN THE predawn hours, we piled into the car and headed for the security gate. Caravanning alongside us was a local farmer named Baek Jong-han, who maintains a small plot of land inside the Cheorwon CCZ. His plot was much more developed than most, with long plastic greenhouses that shielded his crops from direct sunlight. As the morning's bright orange glow crept up the back side of the mountains, we walked into one of his makeshift barns, which was filled with row after row of red chilis drying in the crisp air. Cut into the far side of the greenhouse was a black rectangle with a shoestring pulley. As we approached, the sound of crane calls became deafening.

When the farmer and I approached the wall, he pulled gently on the shoestring, which opened a small window, providing a direct view of hundreds of white-naped cranes roosting together in a semipermanent city atop a frozen lake. The birds were loudly chattering away, perhaps discussing their plans for the day's foraging, before bidding adieu to their neighbors and jetting off for the fields in groups of twos, threes, and fours.

To get an optimal view of where these families were headed next, we hopped back in the car and drove up the side of Sabseulbong, a prominent mound that rises high above the otherwise flat basin.

At the height of warfare in the summer of 1951, this small mountain, which provides highly advantageous, 360-degree views of the region, was among the most fiercely contested spots on the entire peninsula. As the South Koreans fought alongside UN forces to secure the area near the 38th parallel, they were frustrated to continually get pushed back into a stalemate position where neither side could advance any farther. In desperate need of a win, an order was issued from on high: Recapture Sabseulbong at all costs.

As part of the strike, UN forces fired more than two hundred artillery guns in unison, spilling over two hundred thousand shells. More than seven hundred bombs were dropped from aircraft overhead, spewing heavy smoke and dirt throughout the region. For seven days and nights, possession of the mountain changed hands, back and forth, sometimes several times a day. By the week's end, twenty thousand people had been killed or injured in the bloodbath. So complete was the mountain's mutilation that not a single tree remained, and the charred, convulsing earth started to melt down the hillside, giving rise to the mound's inglorious nickname: Ice Cream Hill.

This mountain carries a personal significance for George, I would learn. When he visited Korea in 1977, he and his research team set about visiting every place they could from the west coast to the east. At each of the twenty-two military posts they stopped at along the way, they would wake up early in the morning and listen intently for the sound of distant calls, almost always hearing none. He spotted a few dozen red-crowned and white-naped cranes near Panmunjom in the west, but that was followed by tallies of zero, zero, zero at every stop between there and Cheorwon. By the time his team was halfway to the sea, the mood was grim.

There was no road or footpath to the mountain in those days, he said, and the rice fields around it had not yet been formally reclaimed, so he had to trek overland. In the months leading up to his arrival, South Korean soldiers had excavated the soil around the hill's base and discovered a mass grave with thousands of bodies.

One cold, overcast day, he used his telescope to scan the horizon from this very peak, looking for signs of life. When the morning clouds shifted, his eye spotted what appeared to be a piece of white plastic in the distance. Then, the plastic started moving, flapping its wings. Zooming out, he saw that this bird was not alone. As the fog lifted like a curtain, it revealed a field full of cranes, including three white-naped and at least 125 red-crowned.

Grinning broadly as he reflected on this memory, George said he was convinced that the cranes were as integral to the basin's rebirth as the sprouting trees he's watched grow year after year. Today, Ice Cream Hill has recovered, standing tall and wearing what government signage describes as fresh "green clothes" as it stares at North Korea "without a word."

Since wild cranes live an average of twenty to thirty years, the birds George spotted that foggy day almost five decades ago are no longer with us—but their children, grandchildren, and great-grandchildren are. For them, as for him, this remains hallowed ground.

The good word also appears to have spread throughout the global crane diaspora. Like many animals, cranes strive to reach a group consensus on major decisions like migration routes and wintering destinations. They are also adaptive learners, acquiring knowledge from past experiences, which informs their future decision-making.

Yuko Haraguchi, curator at Crane Park Izumi in southern Japan, which has historically served as the end of the line for the eastern flyway, told me that many cranes who once spent time in the DMZ before continuing on to Japan have stopped coming farther south in recent years. They have, it appears, decided that the Cheorwon Basin, with its abundant food, absence of perceived threats, and clear sense of community, is their promised land.

Today, sandwiched between the old guard barracks atop Ice Cream Hill, stands the wooden viewing platform of a shiny new crane observatory. From this breathtaking vantage point, George can now behold a vast plateau populated every winter by up to one thousand red-crowned and ten thousand white-naped cranes—the largest congregation of its kind on Earth.

AS UNIVERSAL SYMBOLS of peace, cranes are often viewed as mediators of a sort. Plumed sages who occupy fraught spaces to call upon others to listen to their better angels. But these gallant birds are no more in our service than the beavers who tame cascading streams, or the mother bats who soar to great fanfare into the summer sky.

Whether motivated by personal ambition or simply a desire for safe refuge, they inhabit places that have meaning and value for *them*, irrespective of our histories and associations.

The inadvertent circumstances that turned the DMZ into an avian Elysium make it a flawed template, but it remains a particularly vivid example of how deeply a physical place can penetrate the hearts of its nonhuman denizens. It also demonstrates what's possible when animal populations are afforded the latitude to thrive.

Moving forward, any just development of the Cheorwon Basin will inevitably be a trilateral negotiation between North, South, and crane. It's no longer possible to change this place without displacing someone. Given the interspecies language barrier, experienced translators and cultural emissaries like George may be needed to articulate what exactly the birds require. However, in many cases, what other animals need from us is not proactive support, but rather informed deference.

The cranes of the Korean peninsula, like each of the distinctive animal nations I've had the honor of visiting, have their own nuanced hopes and dreams. But there are constants. The need for shelter and nourishment. The desire to live free from harm and the aspiration to raise their young in places they deem secure. The pathways they follow to achieve these goals differ, though in the end, they want what we want.

Every day, as new knowledge emerges about other animals' sophisticated societies, the world we thought we knew fades further into the recesses of our memory—an incomplete history superseded by new stories we get to write.

Through that evolution, we have the ability to continuously reshape how we interact with the nonhuman populations of the world. Gestures great and small, which facilitate stronger cohesion where our needs and desires overlap. And all the while, our fellow Earthlings will update their own worldviews in reaction to us.

Responding effectively to new information is, after all, a fundamental building block of every great animal society. It offers a chance to more fully appreciate the rich diversity of nations as we push together in the same direction, with shared ambitions and purpose, until our journey's end.

Epilogue

RIFLING THROUGH A thick stack of hand-scrawled notes, USB drives, and marked-up maps in my home office, I couldn't shake the feeling that something was off. Seated at my desk on a dim January morning, the epiphanies were not yet flowing to the written page. From my notebook, I read scribbled phrases aloud to transport myself back to certain moments. "They eat the wild ginger between the crops!" "Catalonia is *disputed* territory." "Brush, brush, more brush, sudden clearing, lodge!" I pinned to the wall a small flag that George had gifted me upon my departure, inscribed "To Ryan—With warm memories from the cold Korean DMZ."

But for all the memorabilia, distractions were getting the better of me. Maybe a fresh cup of tea will get the creative juices flowing, I thought, or a late-morning snack. Oh, and how long has that mole-rat email been sitting in my inbox? I should reply to that.

I spun on my chair, waiting for inspiration to strike, until it ran out of momentum facing the corner windows. Beyond them, the mountains appeared as a blurred silhouette through the morning fog. Of course, I thought. That's where the words come from.

I quickly packed up my things, jumped in the car, and cruised up Route 39, which cuts through the foothill city of Azusa before weaving its way through a series of deep, wildfire-singed valleys. I rolled past the Morris and San Gabriel Reservoirs and into the heart of Angeles National Forest, winding two, three, four, five thousand feet upward as the seasons played in fast-forward out my window. In less than an hour, palm trees gave way to sagebrush, only to be overtaken by snow-dusted coniferous pines that cloaked the road in gloomy shadows—a microclimate teleportation that is, to my mind, the most sublime feature of Californian ecology.

As I drove, my mind began to kindle, reliving the past half year's slideshow. Floating above the fog and smog, I was temporarily transported by the pale blue skies to sweltering evenings in the Texas Hill Country. Hugging the byway curves recalled the exhilaration of zipping up and over Corsica to reach the Argentine ants' sandy fortress on the far side of the hills. My memories were not of visits to wild places where animals happen to live, but to "animal places" wholly shaped and defined by the nonhuman peoples who inhabit them.

I drove for another few minutes, waiting for just the right pull-out before parking on an unnamed ridgeline that overlooked a steep valley. Despite the bitter temperature at this altitude, I channeled my inner Kisup, rolled the windows down, and started writing.

For the next six months, the woods in and around Greater Los Angeles were where I turned whenever I hit a stumbling block. I jaunted up every forest road I could find. Angeles Crest Scenic Byway, Upper Big Tujunga Canyon Road, Chaney Trail, Mount Wilson Red Box Road, Glendora Mountain Road—always a different parking spot. A good portion of this book was written from the roadside, and no chapter was complete without a full read in the open air.

Leaving home and manifesting memories of the hidden nations I traversed allowed me to reaffirm that my prose felt true not only to my own experiences, but to the identities of the places themselves. I sought to imagine, as much as any of us can, what life in such a society might be like from the animals' perspective. These stories, after all, are not merely my own.

As I swerved back and forth across the wildland-urban interface, the implicit separation between my world and the forest behind it began to erode. The ease with which I darted to and from the woods was itself a demonstration of humanity's nonzero imprint on this land. But in a new and visceral way, I also started to view the day-tripping bears, deer, and coyotes who wander down my neighborhood's narrow canyon streets through a similar lens. I became attuned to the fact that for each of us, this pulsating region is one of rich gradations. A technicolor tapestry where the spaces in between my destinations are landmarks for others. Where human and nonhuman populations ebb and flow, but rarely is anyone truly alone.

In time, trips to the wilderness felt less like leaving civilization behind than entering new ones. Where I once would have experienced the gratification of departure, I felt only the warmth of arrival.

ON A RARE, clear day, one can look due west from the Mount Wilson Observatory parking lot, perched at nearly 6,000 feet directly above my cabin, and see the wide basin of the San Fernando Valley. A natural bowl flanked by mountains, the valley floor once teemed with life, as foxes, hares, deer, and coyotes coursed through its trough, occasionally plucked by skulking cougars, also known as mountain lions.

For much of that time, this valley also served as a primary transit corridor for the Chumash people, who traveled between inland fields and the coast to gather acorns and seeds, as well as to trade with the Tongva and other Native tribes. As in so many places, the trails these tribes followed were invariably based on the tracks of elk and other nonhuman animals, who had already discovered the most efficient routes from place to place.

When Spanish colonists arrived in the eighteenth century, they, too, exploited many of these well-established human-animal paths, redesignating trodden dirt as the southern leg of what they christened El Camino Real—the Royal Road. Often referred to as California's first highway, it connected twenty-one Franciscan missions from San Diego to Sonoma and formed the spine of what would later become the 101 freeway, a ten-lane wall of speeding steel that bisected the southwest corner of the Valley and obliterated its natural landbridge. When that concrete river and its tributaries were poured, it fragmented a Pangean region and marooned animal populations on an archipelago of newly birthed terrestrial islands.

Without a viable path across one of the busiest roadways in the world, the cougar population trapped on the south side of the 101 now has among the lowest genetic diversity ever recorded, a result of unavoidable inbreeding between brother and sister, or parent and child, triggering what researchers call an "extinction vortex."

In their desperation to break the cycle, some intrepid individuals make foolhardy attempts to leap between the earthen isles by sprinting across the freeway. One cool, late-spring morning, under cover of the marine layer, a young, unidentified cougar tried to cross the divide, but didn't get far. At 4:48 a.m., the California Highway Patrol received a call about a deceased animal in the southernmost lane, near the intersection with Liberty Canyon Road.

When I heard the news, all I could think was, "If only he'd held out a little bit longer."

I had, in fact, made several recent visits to that exact intersection, so I instantly understood why it was the one the cougar chose.

It's the region's only remaining patch of highway with enough protected space on both its north and south slopes to facilitate a discreet approach to the road's edge, and countless animals have been photographed in the same spot as they look longingly to the other side.

From the hills to the north, cougars can glimpse the expansive Santa Monica Mountains, one of only two "biodiversity hotspots" left in the continental United States. Under a business-as-usual scenario, the looming extinction vortex could leave those mountains completely cougar-free within fifteen years.

From atop the Phantom Trail, which skirts the steep hills along the freeway's south side, hikers are rewarded with a transfixing view of the Simi Hills and Santa Susana Mountains. Beyond them, the 2,700 square miles of Los Padres National Forest beckon.

During one of my nighttime visits to the trail, phantasmal silhouettes darted through nearby bushes as midnight approached, illuminated ever so slightly by large spotlights down on the road. Then, at precisely 11:36 p.m., I watched as northbound traffic was abruptly rerouted onto an adjacent access road and thundering tire rumbles were temporarily replaced by the gentle beeps of concrete mixers and construction cranes lowering girders into place.

I was drawn to this particular intersection because it's undergoing an expansion, but not to build more lanes or on-ramps. Not in the traditional sense, anyway. In fact, once it's done, the new addition won't be paved at all. Rather, it will be landscaped with a football field's worth of coastal sage scrub, toyon, native grapes, sunflowers, oak trees, and dozens of other native plants.

Once those transplanted roots dig in, Liberty Canyon will be home to the largest wildlife crossing anywhere in the world. A mammoth 210-foot "vegetated bridge" that will, at least at this one critical choke point, seek to restore a natural link relied upon for millennia by not only mountain lions, but also coyotes, bobcats, birds, frogs, butterflies, and lizards.

AIR CRANE

Returning to the area for the first time since the nameless young cougar met his demise, I arranged to join Jewlya Samaniego, an ethnobotanist of Chumash and Tataviam ancestry, for a tour of the purpose-built nursery she maintains a stone's throw away from the construction site. Inside, she showed off some of the six thousand plants she was growing as part of phase 1 of the project, all of which were cultivated from seeds she personally foraged within a seven-mile radius. By the year's end, she planned to relocate them all to the crossing site and start fresh with another forty thousand plants as part of phase 2, which includes the on-ramps that connect the crossing to the adjacent hillsides. Once the phase 2 plants are installed, her job is complete and nature takes it from there.

If her efforts are successful, Samaniego said, the crossing structure will be so seamlessly integrated into the hills that it "will feel like it was always there." A space reinvigorated by the same thick greenery that graced this place when her ancestors foraged across it centuries earlier.

The wildlife crossing, she suggested, is a good-faith attempt by humanity to leverage ecosystem engineering to turn back the clock, deconflict our earlier intrusions, and acknowledge that this place has always been a vital crossroads for far more than one species. A tangible affirmation of the sovereignty of animal nations.

Unwinding the impacts of one of the world's busiest highways is no easy feat. More than three hundred thousand cars pass through Liberty Canyon every day, according to Lauren Gill, a deputy director of the National Wildlife Federation (NWF). That congestion creates extensive noise and light pollution, meaning the new crossing will need soundproof walls to reassure animals that it's safe to pass. All told, this "ecological stitch" will require more than 2,700 tons of steel and 39,000 tons of concrete to once again knit together two biological communities. There is also talk of building a separate viewing platform on the hillside overlooking the crossing, in response to intense public curiosity.

But for Samaniego, completion of the project carries a higher significance. In the Chumash tradition, the first people on Earth were believed to possess both human and nonhuman traits. That all changed when a great flood swept across the land, prompting many people to transmute into a fully animal form to survive. To this day, Chumash folktales often begin with "When the animals were people . . ."

Many of those "animal relatives," she believes, carry the wisdom of her elders. Coyotes, for example, are revered for their cleverness. Raccoons are viewed as mentors who teach young people about which local plants are edible. In a very real sense, she says, by allowing wild animals to move freely again, the crossing and the tracking data that underpins it enable us to "listen to the animals and hear what they have to say."

THE LIBERTY CANYON crossing is the boldest entry yet in a raft of projects that seek to reverse disruptions inadvertently created while building our own infrastructure.

Just twenty minutes farther up the road, Gill from NWF showed me around the Conejo Grade, where the 101 tumbles down from the mountains and onward to the coast. GPS tracking and camera-trap data obtained by the National Park Service show that cougars, gray foxes, bobcats, mule deer, and other animals will often walk right up to the highway at this point, she said, but rarely attempt to cross it due to the near certainty of being struck by a car. It's one of hundreds of locations her team has identified where vegetated bridges and other crossings could reconnect wildlife communities across the state. If their efforts are successful, Gill says, a cougar could theoretically travel from the Southern California coast to the Sierra Nevada—hundreds of miles to the northeast—without touching concrete once.

Looking at NWF's statewide connectivity plan, I flashed back to the multispecies campus map that Dr. Sarah Crowley had previewed for me at Penryn a year earlier, and the "right to roam" concept that underpinned it. Her ambitious vision of shared spaces that all species could seamlessly crisscross appeared to be taking shape on a grand scale in California.

Vegetated bridges serve as multifunctional spaces used by all sorts of animals, but sometimes crossings are designed with highly specific users in mind. In Yosemite's Sonora Pass, NWF identified an ideal place to insert a slatted steel grate into State Route 108, which now allows toads to successfully cross the road just inches below the rolling tires overhead. Up in wine country, NWF plans to build a crossing for the thousands of newts whose migration out of a rural lake was interrupted by Chileno Valley Road. As a temporary measure until construction is completed, newts have been hand carried across the road in buckets by a group of volunteers calling themselves the Chileno Valley Newt Brigade.

In some cases, facilitating such reconnections requires only minimal investment and the simplest of constructions. On a visit to an underpass below the 118 freeway, Gill showed me how the decades-old road had been intentionally built with a drainage culvert to allow rainfall to flow underneath it. Unfortunately, this drainage included a 10-foot vertical concrete cliff, which works fine for water, but amounts to a death drop for wildlife. Camera traps affixed to the underside of the road documented only eight detections of animals utilizing the drainage, most of which were raccoon families who carefully pawed their way down a nearby maintenance ladder. But after discussions with NWF, the California Department of Transportation added a simple concrete ramp, which transformed the drop-off into a gradual slope. Once it was added, the same cameras captured more than a thousand wildlife uses, not only by grateful raccoons, but also by coyotes, skunks, and at least one black bear. Total cost of the ramp? About $12,000.

Studies show that animals will go to extraordinary lengths to avoid the perils of our highways, often making early attempts to utilize wildlife crossings before they're even finished. Once construction of an overpass or underpass is completed, it can reduce animal-vehicle collisions by up to 97 percent.

In Utah, more than seven hundred animals now trot over a high-elevation wildlife crossing that spans I-80 each year, including moose, bears, and porcupines. In 2024, Colorado opened the first in a series of wildlife overpasses and underpasses just west of Denver, which provide reliable routes for elk and bobcats. Across the Mountain West, bighorn sheep, pikas, coyotes, pronghorn antelope, and many other native animals are suddenly returning to areas they had been unexpectedly cut off from. Along the Columbia River in Oregon and Washington, a system of "fish ladders" enables salmon—who sometimes travel up to a thousand miles to feed in the open ocean before returning at the end of their lives to the exact rivers where they were born—to bypass upstream disruptions created by human dams.

Similar stories are playing out around the world. In Belgium, a bridge spans Brussels Ring Road to facilitate free transit between the surrounding forests by foxes, amphibians, and even insects. Wild boars in Germany, lynxes in the Czech Republic, red squirrels in the UK, tigers in India, koalas and possums in Australia, and macaques in Singapore have all benefited from specially built structures that empower them to get where they need to go. In perhaps the most striking example of all, a steel bridge over Murray Road on Christmas Island in the Indian Ocean—human population: 2,000—allows safe passage for millions of red crabs to get from the forest to the ocean.

The proliferation of such benevolent structures stands as proof that, when we want to, we can do great things. They send an intelligible but soft-spoken message from humanity to the rest of the animal kingdom that says, "We see you."

I returned to my car and pulled back onto the freeway, allowing my mind to wander, contemplating how far we might be able to take this transformative concept. I pondered high-wire bridges that span between treetops so animals can traverse the arboreal gaps above our streets. Narrow subterranean tunnels that wend beneath gritty urban intersections. Multispecies roads that could be unobtrusively shared by humans at times of the year when seasonal migrations take wild animals elsewhere. And given the rapid pace with which we are collecting new data about areas important to other animals,

I mused about some sort of notification system that allows human travelers to know when they're passing through a place densely populated by another species.

In that moment, nothing felt impossible.

REJUVENATED BY ALL I'd seen, I began to feel the limits of my own home range fade away. Adrift in a fertile place, my attention turned from familiar day-tripping haunts to the fuzzy fringes. Fixed coordinates once settled at the end of the road now anchored jumping-off points into unbounded spaces, where true illumination awaits.

Barely half an hour beyond Conejo Grade, following the mission bells of the former El Camino Real, the 101 crashes into the coast before taking a sharp right and hugging the sand all the way to Santa Barbara Harbor. After parking at the edge of western civilization, I unloaded my dry bag and waterproof notepads—purchased more than a year earlier for the Canadian muskeg—and met Captain Garrick Hyder for a sunrise departure.

From the deck of his small liveaboard boat, Hyder's front-porch view includes the hulking silhouette of Santa Cruz Island, which tempts mainland vacationers with its sea caves and rugged hiking trails. The largest and closest of California's four northern Channel Islands, it is flanked on either side by Anacapa and Santa Rosa, both less visited but perfectly accessible.

Today's destination, the remote, westernmost island, known as San Miguel, remains far out of sight, ensconced in thick fog, 43 miles from the beach. "As a charter boat captain, I've taken hundreds, maybe thousands, of people to Santa Cruz, Anacapa, and Santa Rosa," Hyder said. "You're the first person to ever hire me to go to Miguel."

Called Tuqan by the Chumash, who maintained a number of

permanent villages that seafaring residents reached from the mainland in redwood canoes known as *tomols*, San Miguel has been "uninhabited" since the 1940s.

As we pushed off and the ocean floor began to drop beneath our feet, an entirely different universe opened up. Astronomers continue to optimistically scan the Great Beyond looking for signs of alien civilizations, but here on Earth we have only skimmed the surface in searching for intelligent life beneath the waves.

Once presumed to be largely empty, the bottom of the Pacific—parts of which are deeper than Mount Everest is tall—is now understood to be home to all manner of mysterious creatures, from deep-sea fish and crabs to giant tube worms more than nine feet long. And those are just the ones we know about. When scientists conducted a series of expeditions to the bottom of the Clarion-Clipperton Zone, a vast stretch of open ocean between Hawaii and Mexico where mineral deposits line the seafloor, the researchers identified at least five thousand previously unknown species. On a global level, the National Oceanic and Atmospheric Administration (NOAA) says that more than 80 percent of the ocean remains "unmapped, unobserved, and unexplored," but is nonetheless believed to contain upward of a million species, at least two-thirds of which we've never seen before.

Even the animals we do know about continue to surprise us. In 2009, scientists in Jervis Bay, off the eastern coast of Australia, were stunned to spot a large group of gloomy octopuses,* who were widely understood to live solitary lives. At least until the researchers stumbled onto a cluster of them congregating, communicating, and dwelling together, sometimes in groups as large as sixteen individuals, in an area later dubbed "Octopolis." Octopolis was initially assumed to be an anomaly, until eight years later when a separate site containing another fifteen octopuses was discovered nearby, charm-

* Not a personal slight: "Gloomy octopus" is the common name for this species.

ingly christened "Octlantis." It's unclear how widespread such practices are, but experts say that it's probably not the octopuses who are changing their behavior. We're just getting a late start on learning what they're up to.

"Everything is more challenging in the marine environment," NOAA ecologist Marc Lammers told me. Since solar panels are of no use underwater, GPS trackers can run only as long as battery power allows. Radio transmitters can be used to monitor larger sea mammals, but researchers have to wait for them to come to the surface before the signal comes through. Acoustics can be useful to triangulate some animal movements, since sound transmits four and a half times faster in water than in air. By using recorders to listen to ocean chatter and then feeding those recordings into a machine-learning tool that can separate out specific sounds, like beluga whale calls, we can start to make sense of who is passing through. But that works only for animals who make sounds, Lammers says. "Those who don't are largely invisible to us."

COMFORTABLY CRUISING ALONG the upper edge of the submerged cosmos, Captain Hyder was feeling validated in his decision to delay our previously scheduled trip to Miguel by several weeks. As the first place where pummeling winds make landfall after screaming across the open ocean, the island is often ringed by eight-foot waves that could easily capsize a boat like his. But today, the sea was as close to glassy as it gets, gently swelling like the chest of a sleeping giant.

The bountiful waters surrounding the island are protected by the Channel Islands National Marine Sanctuary, part of a network that has so far designated more than 600,000 square miles as being of particular importance to marine life. Just north of San Miguel, the Chumash also put forward an ambitious sanctuary proposal—the first ever nominated by a Native tribe—which now protects an additional 4,543 square miles in an area that has been called the "Serengeti of the Sea." Collectively, America's marine sanctuaries offer increased protection to everyone from sea turtles in Georgia to humpback whales in Hawaii, some of whom travel up to 5,000 nautical miles round-trip to visit the same tropical waters every winter.

Within designated sanctuaries, many disruptive activities like drilling and mining are restricted or prohibited completely. Some also have speed limits that carry strict penalties to minimize the risk of boats colliding with marine-animal residents. Here in the Santa Barbara Channel, national sanctuaries even offer trophies to the shipping companies that move the slowest.

All along the two-hour journey to the island, our boat was chased by pods of mischievous dolphins. "Most marine mammals are terrified of boats, but these dolphins will go out of their way to follow us like dogs chasing cars," Hyder said. "You see the giddy look on their faces and you can tell they're excited."

Around the halfway point, a massive humpback whale briefly crested and paused on the surface before taking a deep dive. As he arched his mammoth tail, roughly the size of a Buick, we glimpsed the signature lobes and scalloped edges that make each one as unique as a human fingerprint. They're so distinctive that researchers have managed to train machine-learning tools to recognize specific whales simply by uploading a tail photo. More than one hundred thousand individuals have already been conclusively identified, helping us to make better sense of who goes where.

As San Miguel's outline became faintly visible through the marine layer and the winds started to pick up speed, Hyder repeatedly refreshed his live weather and sea trackers to gauge the best strategy for an approach. Scanning across a screen cluttered with squiggly red warnings about unclassified restricted zones, he zoomed in to navigate between marked danger areas "to be avoided."

To Hyder's delight, the far western corner of the island, known as Point Bennett, appeared to be in play. "That place is mythical," he said. The ferocious winds that regularly slam into its shores are legendary for pushing the relatively warm surface water away, driving an upwelling effect that thrusts cold, nutrient-rich waters up from the deep. "You've never seen more fish."

Given its intense natural conditions, Point Bennett is also one of several places on San Miguel renowned for its shipwrecks. In 1923, a 307-foot cargo passenger ship called the *Cuba* got lost in fog off San Miguel for three days before striking a reef near Point Bennett, where part of the wreck remains to this day.

About 95 percent of the time Point Bennett is far too exposed to the elements for small boats like ours to enter. But this morning a brief window of calm appeared to be narrowly opening the gates for us.

As we heaved into the zone, Hyder noted that the relative tranquility of the waves belied the reality beneath. "This is one of the sharkiest places in the world. I call it 'the lunch line.' There are absolutely sharks with us here right now, and they're not the small ones. They're the *big* ones." Three decades ago, he said, it was in this exact spot that a commercial urchin diver named James Robinson was infamously ripped apart by a great white more than 16 feet long. His death was so immediate that a doctor reportedly said, "If he'd been bitten on the steps of the hospital, I still couldn't have saved him." Hyder himself does a fair amount of deep-sea diving, and he considers very few spots along the West Coast off-limits. But Point Bennett is one of them. "This place," he noted with a rueful grin, "scares the shit out of me."

Once we carefully cleared the western corner and turned toward

San Miguel's leeward side, a whole new set of sights and sounds came into view as hundreds of sunbathing seals and sea lions waddled up and down an unmarked beach. Chattering colonies barked, moaned, and grunted to each other. It sounded as though we'd suddenly arrived on a mystical island of Chewbaccas.

San Miguel falls in a special zone where cold currents moving down the Pacific coast intermingle with warm currents flowing up from Mexico. This confluence has turned the island into an aquatic crossroads of sorts, where it serves as both a northern terminus for transitory seal populations from Mexico and a southern terminus for northern fur seals, some of whom have traveled all the way from the Pribilof Islands in Alaska.

In total, the National Park Service estimates that at least 1,100

harbor seals, 5,000 northern fur seals, 50,000 northern elephant seals, and 70,000 California sea lions—including a whopping 45 percent of the global breeding population—visit San Miguel each year. The pinnipeds spend the vast majority of their lives traveling and foraging in the ocean, but every day, some portion will lurch up from the sea to unwind, socialize, molt, give birth, or nurse their children on its white-sand beaches.

Sheltered from the wind, Hyder and I pulled into a nearby cove filled with lush kelp beds, shut off the engine, and quietly absorbed the remarkable scene. Lingering a few hundred yards from the beach, we watched as playful groups of sea lion pups called out to their mothers and skidded down the pebbled hillside. Blubbery males as-

sertively slapped the ground and barked at one another, as a rotating cast of colorful characters came and went between the sand and sea.

Given that each species uses distinct vocalizations and behavior to express themselves, communications between the various groups are generally limited to obvious territorial displays and "leave me alone" signals, which to my eye gave the panorama the vibe of a congested street corner. Close contacts engaged in vigorous conversation, surrounded by the cross talk of anonymous passersby.

Rolling, cerulean waves carried our small boat along San Miguel's southern coast as we slipped past Judith Rock, across Tyler Bight, and beyond Crook Point, with pinnipeds inhabiting every sandy stretch and rocky outcropping along the way. Many thousands in total. Upon returning to the east side, we completed our circumnavigation by anchoring in Cuyler Harbor, once home to the isle's largest Chumash village, now a sparse beach that could reasonably accompany the dictionary definition of a "desert island." Tossing two paddleboards into the water, the captain and I briskly propelled onto its wind-scarred shore, making landfall near three exhausted palm trees.

On the hillside above hung the shattered remains of a wooden staircase, which once led to the island's nine official campsites until a winter storm caused a landslide that closed the route indefinitely. This year, Hyder said, the number of human visitors to this beach could probably be counted on one hand.

Reflecting on the surreality of it all, I looked down at the sand and noticed an uninterrupted chain of footprints from one of the island's many seabirds. San Miguel, Hyder noted, is also believed to be home to a number of terrestrial animals, from deer mice to reptiles. Some, like a unique subspecies of island fox the size of a house cat, exist nowhere else. For them, as now for me, this distinctive island remains anything but deserted.

As I stood on that foggy beach, savoring the final moments

before lowering my paddleboard back into the sea, I felt the same familiar warmth I'd come to expect upon entering the forest. A radiant, sustaining energy similar to what Beston described feeling as he came to appreciate the true abundance that exists around us:

> There were times, on the threshold of spring, when the force seemed as real as heat from the sun. A sceptic may smile and ask me to come to his laboratory and demonstrate; he may talk as he will of the secret workings of my own isolated and uninfluenced flesh and blood, but I think that those who have lived in nature, and tried to open their doors rather than close them on her energies, will understand well enough what I mean.

Acknowledgments

As I hope is clear from my story, I did not do this by myself. Given the scope of where I wanted to go and what I wanted to cover, this book was a big swing, made possible through the contributions of hundreds of collaborators—far too many to list.

Instead, I'd like to briefly acknowledge some of the raw elements that made my journey possible.

Amators

More than anything else, this book was powered by people who pursue their area of interest not for financial gain but because they *absolutely love it*. The scientists, historians, artists, and laypeople so passionate about their niche topic that they jumped at the opportunity to talk for hours with a complete stranger.

When researching a new subject, there is no better feeling than stumbling upon someone who stays up late in the evenings, quietly connecting with a small pool of like-minded peers and publishing reports, maps, or blogs to share their latest discoveries. People for whom a particular corner of our planet or the animal kingdom has proved so entrancing that it absorbs them completely. It was an honor to get even a small glimpse into how big the world looks through their eyes.

Open-access information

YEARS BEFORE ANY publisher agreed to take a chance on this book, it was a concept that occupied my mind in the predawn hours before daily life began. A passion project I pursued primarily by scouring digital libraries and dispatching chaser emails to experts I hoped could help me better understand what we know, what we don't, and where we might be going next.

I was able to do so only because of the wide array of entities for which the widespread dissemination of knowledge is the only goal—the academic institutions, government agencies, and nonprofit organizations committed to releasing their insights free of charge, for curious onlookers like me to peruse. Many of the individual researchers I am most deeply indebted to have been cited in my endnotes so that others might similarly benefit from their learnings. But for every person I've cited, there were ten others whose off-the-record conversations helped further shape the finished product.

I am particularly grateful to those interviewees who didn't end our discussions when we exhausted the depths of current knowledge, but rather leaned back and ruminated with me about how to describe and depict the innumerable things we do not yet understand about the animal kingdom and humanity's role in it. Some of this idle speculation made its way into the book, with the appropriate caveats, and will almost certainly be proved true in due time.

Flexibility

AS A RULE, expeditions into unknown lands do not adhere to fixed schedules. They are born from a steadfast commitment to follow the story wherever it leads, which requires constant reexamination and adjustment by the traveler and everyone in their orbit. It's an indulgent pursuit, contingent on the kindness of strangers and inexhaustible patience of friends, colleagues, and loved ones, which I received in great abundance—most notably from my wife, Meghan, and daughter, Nellie.

No man is an island, but those of us who wander are particularly dependent on the vitality, deftness, and patronage of everyone they meet along the way.

And finally, first drafts

I APPROACHED THIS book with the full knowledge that it is the beginning of a conversation, not the end. Just as early maps provide, at best, a partial picture of reality, exploration is, by its very nature, incomplete. An honest attempt to describe an unfamiliar place that will inevitably be expanded, refined, corrected, or wholly replaced as new perspectives emerge. As Catalonia's King of Ants rightly noted, each of us is limited in time, no better or worse than those who will succeed us. An equal. "Just one."

In that sense, my story is also reflective of where we find ourselves as a species: a few steps beyond the starting point en route to a destination we cannot yet see. Staring off into the shadowed woods ahead, one could be forgiven for momentarily thinking that our journey will be a solitary trek—but by now we know better than that.

Notes

All websites were live as of December 2024. All phone, email, virtual, and in-person interviews were conducted by the author.

Cartographic sources: To build the contextual layers underpinning our maps (roads, rivers, borders, and so on), we used OpenStreetMap and Natural Earth. In addition, we obtained land use data from ESA WorldCover, terrain data from NASA's Shuttle Radar Topography Mission (SRTM), and bathymetric data from the General Bathymetric Chart of the Oceans (GEBCO). Other map sources are listed in specific endnotes.

PROLOGUE

1 **sent to Cape Cod:** Don Wilding, Cape Cod and Henry Beston historian and author, email, July 8, 2024.

1 **three thousand shipwrecks:** "Lifesavers," Cape Cod National Seashore," National Park Service, *https://www.nps.gov/caco/learn/historyculture/lifesavers.htm*.

2 **"We patronize them":** Henry Beston, *The Outermost House* (Penguin Books, 1988), 25 (emphasis added).

3 **"no homes have been destroyed":** KCAL News, "Fresh Evacuation Issued as Stubborn Bobcat Fire Containment Delayed," CBS News, September 17, 2020, *https://www.cbsnews.com/losangeles/news/bobcat-fire-evacuations-antelope-valley/*.

3 **caused "no injuries":** Hayley Smith and Louis Sahagún, "How the Bobcat Fire Grew into a Unique Menace That's Evading Firefighters," *Los Angeles Times*, September 15, 2020, *https://www.latimes.com/california/story/2020-09-15/elusive-bobcat-fire-became-a-dangerous-menace-without-burning-any-homes*.

3 **rabbits and squirrels to coyotes and mountain lions:** Tim Daly, public information officer, California Department of Fish and Wildlife, email, September 11, 2024.

3 **Her name is Elsie Eiler:** Alexx Altman-Devilbiss, "Population 1: 88-Year-Old Woman Explains What It's Like Being the Only Person in Town," National Desk, July 18, 2023, *https://cbsaustin.com/news/offbeat/population-1-88-year-old woman-explains-what-its-like-being-the-only-person-in-town*.

6 **first institution in the world:** "Up on the Roof," University of Exeter, May 15, 2015, *https://news-archive.exeter.ac.uk/research/esi/2015/articles/upontheroof.php*.

6 **new species of jumping spider:** Patrick Barkham, "Exotic Spiders Flourishing in Britain as New Jumping Species Found in Cornwall," *The Guardian*, April 26, 2024, *https://www.theguardian.com/environment/2024/apr/26/new-species-of-jumping-spider-found-on-university-campus-in-cornwall*.

6 **thirty more individuals:** Dr. Sarah Crowley, senior lecturer in human and animal geography, University of Exeter, email, June 24, 2024.

6 **app called Shell-E:** *https://shelle.cec.ex.ac.uk/*.

7 **Professor Dave Hodgson:** Professor Dave Hodgson, Centre for Ecology and Conservation, University of Exeter, email, July 8, 2024.

7 **ancient Celtic nations:** The Celtic League, *https://www.celticleague.net/*.

7 **miles of public trails:** "Walking in Cornwall," National Trust, *https://www.nationaltrust.org.uk/visit/cornwall/walking*.

8 **[Penryn multispecies map]:** Animal sightings and unpublished pilot data of hedgehog tracks courtesy of Dr. Sarah Crowley, Abhishek Dixit, Dr. David J. Hodgson, Dr. Dave Hudson, Dr. Jonathan Bennie, and Dr. Kevin J. Gaston. The illustration of the Penryn campus was based on imagery from Google Earth, 2024.

11 **[Cumulative studies and animal locations uploaded to Movebank, 2012–25]:** M. Wikelski, S. C. Davidson, R. Kays, Movebank: Archive, Analysis, and Sharing of Animal Movement Data, hosted by the Max Planck

Institute of Animal Behavior, 2025, accessed April 15, 2025, *https://www.movebank.org*.

11 **Movebank adds nearly 12 million:** Sarah Davidson, Movebank data curator, email, August 8, 2024.

11 **squirrels, sloths, parrots, foxes, opossums, sharks, turkeys, flamingos, penguins, and nearly every other group imaginable:** Dr. David La Puma, vice president, global market development, Cellular Tracking Technologies, email, March 16, 2023.

12 **rule of thumb:** Dr. Martin Wikelski, director, Department of Migration, Max Planck Institute of Animal Behavior, email, July 23, 2024.

12 **single grain of rice:** La Puma, email, August 9, 2024.

13 **of the kinkajous' world:** Dr. Meg Crofoot, director, Department for the Ecology of Animal Societies, Max Planck Institute of Animal Behavior, email, August 21, 2024.

14 **Anthony Dancer, a conservation specialist:** Robin McKie, "'Only AI Made It Possible': Scientists Hail Breakthrough in Tracking British Wildlife," *The Guardian*, August 13, 2023, *https://www.theguardian.com/technology/2023/aug/13/only-ai-made-it-possible-scientists-hail-breakthrough-in-tracking-british-wildlife*.

14 **would have taken humans thirteen years:** Dina Dechmann, Benjamin Koger, and Roland Kays, "World's Biggest Bat Colony Gathers in Zambia Every Year: We Used Artificial Intelligence to Count Them," *The Conversation*, September 26, 2023, *https://theconversation.com/worlds-biggest-bat-colony-gathers-in-zambia-every-year-we-used-artificial-intelligence-to-count-them-210028*.

14 **Naik and his collaborators constructed:** Dr. Hemal Naik, postdoctoral researcher, Department for the Ecology of Animal Societies, Max Planck Institute of Animal Behavior, email, December 18, 2024.

15 **into hyper-precise datasets:** Urs Waldmann et al., "3D-MuPPET: 3D Multi-Pigeon Pose Estimation and Tracking," *International Journal of Computer Vision* 132 (May 7, 2024): 4235–52, *https://doi.org/10.1007/s11263-024-02074-y*.

16 **exactly which zebra:** Eric Price et al., "A Framework for Fast, Large-Scale, Semi-Automatic Inference of Animal Behavior from Monocular Videos," *bioRxiv*, August 2, 2023, *https://doi.org/10.1101/2023.07.31.551177*.

16 **well-established transit corridor:** Benjamin Koger et al., "Quantifying the Movement, Behaviour and Environmental Context of Group-Living Animals Using Drones and Computer Vision," *Journal of Animal Ecology* 92, no. 7 (July 1, 2023): 1357–71, *https://doi.org/10.1111/1365-2656.13904*.

17 **their own backyard:** Andrew M. Rogers, Russell Q-Y. Yong, and Matthew H. Holden, "The House of a Thousand Species: The Untapped Potential of Comprehensive Biodiversity Censuses of Urban Properties," *Ecology* 105, no. 2 (February 1, 2024): e4225, *https://doi.org/10.1002/ecy.4225*.

17 **"connect the dots":** Ken Chiacchia, "BirdFlow AI 'Connects the Dots' in Massive, Volunteer Database to Track Migratory Birds," Pittsburgh Supercomputing Center, May 10, 2023, *https://www.psc.edu/birdflow-ai-tracks-migratory-birds/.*

20 **Theroux described maps:** Paul Theroux, *Sunrise with Seamonsters* (Houghton Mifflin, 1985), 281.

1. LIFE ON THE FRONTIER: SETTLEMENTS

23 **"where there are plants one after another":** William Bright, *Native American Place Names of the United States* (University of Oklahoma Press, 2004), 52.

23 **Alberta's oldest human settlement:** "Fort Chipewyan," Regional Municipality of Wood Buffalo, *https://www.rmwb.ca/en/indigenous-and-rural-relations/fort-chipewyan.aspx.*

24 **first mandatory evacuation:** Karen Bartko and Emily Mertz, "Alberta Wildfires: Military Helping Airlift Evacuees out of Fort Chipewyan as Threat Increases," *Global News*, May 31, 2023, *https://globalnews.ca/news/9736102/alberta-wildfires-fort-chipewyan-caf/.*

24 **half the town's fuel supplies:** Jay Telegdi, senior manager intergovernmental affairs, Athabasca Chipewyan First Nation, email, August 9, 2024.

24 **bush planes and boats:** "Fort Chipewyan," Regional Municipality of Wood Buffalo, *https://www.rmwb.ca/en/indigenous-and-rural-relations/fort-chipewyan.aspx.*

24 **1.5 million and counting:** David Wallace-Wells, "There's No Escape from Wildfire Smoke," *The New York Times*, May 17, 2023, *https://www.nytimes.com/2023/05/17/opinion/wildfires-smoke-pollution-distance.html.*

25 **eight hundred Indigenous residents:** Greg Bennett, strategist, Regional Municipality of Wood Buffalo, email, June 24, 2024.

25 **as daily updates:** "Fire Information Archives—Wood Buffalo National Park," Parks Canada, *https://parks.canada.ca/pn-np/nt/woodbuffalo/nature/science_nature/fire_management/feu-foret-fire/archives.*

25 **"We will be going home soon":** Brittany Ekelund and Alex Antoneshyn, "Fort Chipewyan and Area Residents 'Going Home Soon,' Leaders Announce," CTV News, June 18, 2023, *https://www.ctvnews.ca/edmonton/article/fort-chipewyan-and-area-residents-going-home-soon-leaders-announce/.*

25 **a 1,100-mile string:** Ian Frazier, "Deep in the Wilderness, the World's Largest Beaver Dam Endures," *Yale Environment 360*, December 11, 2023, *https://e360.yale.edu/features/worlds-largest-beaver-dam.*

25 **Canada's largest national park:** Tim Gauthier, external relations manager, Southwest NWT Field Unit, Parks Canada, email, June 24, 2024.

26 **visiting Frommer's writer:** "Things to See in Fort Chipewyan," Frommer's, *https://www.frommers.com/destinations/fort-chipewyan/attractions/overview*.

27 **at least one person lives:** "Hierarchy of Settlements," Geography Revision, *https://geography-revision.co.uk/a-level/human/hierarchy-of-settlements/#Settlement_defined*.

28 **not alone in enacting this process:** Greg Anderson, professor of history, The Ohio State University, email, February 3, 2025.

28 **"make sense of the world":** Crofoot, email, August 21, 2024.

28 **"not the custodian of creation":** Anderson, email, June 28, 2024.

29 **a mile thick:** "Ice Age Trail Landscape & Geology," Ice Age Trail Alliance, *https://www.iceagetrail.org/explore/explore-the-trail/ice-age-trail/ice-age-trail-landscape-geology/*.

31 **foundation of their world:** Heidi Perryman, president, Worth a Dam, email, July 9, 2024.

31 **providing defensible space:** "Build a Beaver Dam," National Park Service, *https://www.nps.gov/articles/buildabeaverdam.htm*.

31 **Beavers don't hibernate:** Perryman, email, July 9, 2024.

31 **Neil deGrasse Tyson:** Sam Harris, "Science & Civilization," *Making Sense*, podcast, November 10, 2022, 1 hr., 39 min., timestamp: 34 min., 40 sec., *https://www.samharris.org/podcasts/making-sense-episodes/302-science-civilization*.

32 **150 years later:** David Malakoff, "150-Year-Old Map Reveals That Beaver Dams Can Last Centuries," *Science*, December 4, 2015, *https://www.science.org/content/article/150-year-old-map-reveals-beaver-dams-can-last-centuries*.

32 **125,000 years old:** Dr. Grant Zazula, Yukon paleontologist, Government of Yukon, email, July 22, 2024; and Dr. Natalia Rybczynski, paleobiology research associate, Canadian Museum of Nature, and associate professor, Carleton University, email, August 27, 2024.

32 **maze of navigable canals:** "Beaver Facts," Wild Trout Trust, *https://www.wildtrout.org/content/beaver-facts*.

32 **with man-made creations:** Tom Patterson, former president, North American Cartographic Information Society, email, September 8, 2024.

33 **Muskrats commonly build:** Mary Holland, "Muskrats Constructing Lodges," *Naturally Curious with Mary Holland* (blog), October 8, 2014, *https://naturallycuriouswithmaryholland.wordpress.com/2014/10/08/muskrats-constructing-lodges/*.

33 **sort of multispecies neighborhood:** Kylie McElrath, conservation biologist and muskrat specialist, email, July 2024.

33 **goliath frogs—the world's largest:** Taylor & Francis Group, "World's Largest Frogs Build Their Own Ponds for Their Young," *EurekAlert!*, August 8, 2019, *https://www.eurekalert.org/news-releases/486662*.

33 **larger than humanity's tallest skyscraper is to us:** Predrag B. Slijepčević, *Biocivilisations: A New Look at the Science of Life* (Chelsea Green, 2023), Kindle; appears 23 percent of the way into the book.

33 **Gopher tortoises dig burrows:** Amanda Tower, "Gopher Tortoise: Ecosystem Engineers," Naples Botanical Garden, *https://www.naplesgarden.org/gopher-tortoise-ecosystem-engineers/*.

33 **sociable weaver birds:** Drs. Rita Covas and Claire Doutrelant, project coordinator and project co-coordinator, Sociable Weaver Project, email, July 15, 2024.

33 **own architectural traditions:** Rebecca Dzombak, "These Architects Embrace Local Styles. And They Fly to Work, Too," *The New York Times*, September 9, 2024, *https://www.nytimes.com/2024/08/29/climate/weaver-bird-nests-kalahari-desert.html*.

34 **sheet of loose-leaf paper:** Kristin Carringer, director of enterprise communications, Maxar, email, February 15, 2024.

34 **four inches across:** William J. Broad, "When Eyes in the Sky Start Looking Right at You," *The New York Times*, February 20, 2024, *https://www.nytimes.com/2024/02/20/science/satellites-albedo-privacy.html*.

34 **longest beaver dam:** Jean Thie, "The Longest Beaver Dam in the World," EcoInformatics International, *https://www.geostrategis.com/p_beavers-longestdam.htm*.

35 **[Longest beaver dam aerial image]:** Credit: Parks Canada photo, republished with permission, July 26, 2024.

36 **more than ten thousand years:** Vincent McDermott, "People Arrived in Fort McMurray Wood Buffalo Area at Least 11,000 Years Ago, Finds MacEwan Scientist," *Fort McMurray Today*, August 10, 2022, *https://www.fortmcmurraytoday.com/news/people-arrived-in-fort-mcmurray-wood-buffalo-area-at-least-11000-years-ago-finds-macewan-scientist*.

36 **100 to 400 million beavers:** Luke Burbank, "Beavers: A Rodent Success Story," CBS News, September 16, 2023, *https://www.cbsnews.com/video/beavers-a-rodent-success-story/*.

36 **18 feet tall:** Ilo Hiller, "Beavers—Introducing Mammals to Young Naturalists," Texas Parks & Wildlife, *https://tpwd.texas.gov/publications/nonpwdpubs/introducing_mammals/beavers/*.

36 **Lewis and Clark's westward expedition:** Thomas J. Elpel, president of the Jefferson River Chapter of the Lewis and Clark Trail Heritage Foundation, email, June 18, 2023.

36 **15 feet high and 24 feet thick:** Cristy Stout, *Beaver: Ecosystem Engineers*, Bangor Land Trust, *https://www.bangorlandtrust.org/uploads/8/2/6/3/8263142/beaver_-_ecosystems_engineers.pdf*.

36 **forty individual family lodges:** Hiller, "Beavers—Introducing Mammals to Young Naturalists."

37 **"didn't need to be tagged":** Willi Whiston, "Tracing Roots," Shaw Spot-

light Facebook video, October 28, 2021, 4 min., 19 sec., timestamp: 2 min., 12 sec., *https://www.facebook.com/watch/?v=4757111704307590*.

37 **her book *Beaverland*, citing research by historian Calvin Martin:** Leila Philip, *Beaverland: How One Weird Rodent Made America* (Twelve, 2022), 161–62.

38 **[Fur trader image]:** "Colin Fraser, trader at Fort Chipewyan, sorting his black fox pelts from his lot of 35,000 dollars' worth of fur, between 1890–1900." Source: Library and Archives Canada/Ernest Brown fonds/e011303100-012, public domain image via Edward Roué, reference technician, Access and Services Branch, Library and Archives Canada, August 6, 2024.

39 **"twice as many furs as his canoes would carry":** H. A. Innis, *Peter Pond—Fur Trader and Adventurer* (Irwin & Gordon, 1930), 81, *https://gutenberg.ca/ebooks/innis-pond/innis-pond-00-h.html*.

39 **died a pauper:** Author notes, exhibit during in-person visit to Fort McMurray Heritage Village, May 31, 2023.

39 **richest trading post:** Fort Chipewyan National Historic Site of Canada, Canada's Historic Places, *https://www.historicplaces.ca/en/rep-reg/place-lieu.aspx?id=17705*.

39 **For two hundred years:** "Leave It to Beavers," *Nature*, PBS, May 13, 2014, *https://www.dailymotion.com/video/x5txkzr*.

39 **brink of extinction:** "Beavers," Northwest Power and Conservation Council, *https://www.nwcouncil.org/reports/columbia-river-history/beavers/*.

39 **fifteen separate fires:** Brandon Boyd, fire program manager, Wood Buffalo National Park, email, July 8, 2024.

39 **35,000 gallons of flame retardant:** Josee St-Onge, provincial information officer, Alberta Wildfire, email, June 26, 2024.

39 **South Africa and New Zealand:** Karyn Mulcahy, "How Firefighters from Several Countries Have Been Enlisted to Battle Wildfires in Alberta," CTV News, June 13, 2023, *https://www.ctvnews.ca/edmonton/article/how-firefighters-from-several-countries-have-been-enlisted-to-battle-wildfires-in-alberta/*.

40 **within nine miles:** Boyd, email, September 3, 2024.

40 **scorch up to eighteen miles:** Boyd, email, September 3, 2024.

40 **often try to outrun blazes:** Dr. Erin Bayne, professor, Department of Biological Sciences, University of Alberta, email, June 25, 2024.

40 **bear cubs will instinctively:** Carla Ulrich, "Fort Smith Man Says All Wood Buffalo Fires Should Be Fought. Parks Canada Says It's More Complicated," CBC, June 23, 2023, *https://www.cbc.ca/news/canada/north/wood-buffalo-fire-concerns-1.6887166*.

40 **a dozen years:** Naama Weingarten, "As Wildfires Burn in Alberta Forests, What Happens to the Animals?," CBC, June 3, 2023, *https://www.cbc.ca/news/canada/edmonton/wildfire-alberta-wildlife-forest-1.6858134*.

41 **ten rural cabins:** Telegdi, email, August 9, 2024.

41 **Barely six miles:** Boyd, email, September 3, 2024.

42 **[Map of active wildfire hotspots]:** MODIS Collection 6.1 NRT Hotspot / Active Fire Detections MCD14DL distributed from NASA FIRMS, *https://doi.org/10.5067/FIRMS/MODIS/MCD14DL.NRT.0061*.

42 **turning skies orange:** Jesse O'Neill, "NYC Choked by 'Unhealthy' Heavy Smoke from Canadian Wildfires," *New York Post*, June 6, 2023, *https://nypost.com/2023/06/06/nyc-choked-by-heavy-unhealthy-smoke-from-canadian-wildfires/*.

43 **the thousand-person town:** Chief Marcel Head, Shoal Lake Cree Nation, text message, July 30, 2024.

44 **owned by the Canadian government:** Marina Atkinson, supervisor, North Lands Management, Indigenous Services Canada, email, May 8, 2025.

48 **"sixth Great Lake":** Joseph Shavit, "The 6th Great Lake Was Larger Than All Other Lakes Combined," *The Brighter Side of News*, March 15, 2024, *https://www.thebrighterside.news/post/the-6th-great-lake-was-larger-than-all-other-lakes-combined/*.

48 **more beavers here than ever before:** Jean Thie, president, EcoInformatics International, email, July 12, 2024.

48 **"Mapping Beaver Dams with Machine Learning":** Sarah Derouin, "Mapping Beaver Dams with Machine Learning," *Eos*, June 15, 2023, *https://eos.org/research-spotlights/mapping-beaver-dams-with-machine-learning*.

49 **cul-de-sacs of American suburbs:** Dr. Emily Fairfax, assistant professor of geography, University of Minnesota, video call, July 24, 2024.

49 **98.5 percent accuracy rate:** Fairfax, video call, July 24, 2024.

49 **at least two thousand families:** Fairfax, video call, July 24, 2024.

50 **[Map of beaver dams near Pakwaw Lake]:** Dams: Courtesy of the Fairfax Lab, which used EEAGER to analyze select areas around Pakwaw Lake and then conducted the manual tracing that appears in this map based on high-resolution satellite imagery. Pakwaw Lake population: Head. Beaver density: Thie. Satellite image: Copernicus Sentinel data (2024). Retrieved from Copernicus Data Space Ecosystem Browser (May 10, 2024), processed by the European Space Agency, *https://tinyurl.com/4nthp97n*.

2. URBAN FLIGHT: CITIES

55 **South ends and the Southwest begins:** "Texas Hill Country," Wikipedia, *https://en.wikipedia.org/wiki/Texas_Hill_Country*.

55 **engaged in a territorial dispute:** Kristen Cabrera and Shelly Brisbin, "'Violence in the Hill Country' Chronicles a Time of Lawlessness and

Death in Central Texas," *Texas Standard*, August 17, 2021, *https://www.texasstandard.org/stories/violence-in-the-hill-country-chronicles-a-time-of-lawlessness-and-death-in-central-texas/*.

56 **defines a city:** "City," Merriam-Webster.com, *https://www.merriam-webster.com/dictionary/city*.

56 **certain community advantages:** Emily Shepard, professor in behavioural ecology, Swansea University, email, February 14, 2024.

56 **had a cathedral:** "City Status in the United Kingdom," Wikipedia, *https://en.wikipedia.org/wiki/City_status_in_the_United_Kingdom*.

56 **reasons for coming together:** Shepard, email, February 14, 2024.

57 **nearly 20 million:** Jack Morgan, "Batnado at Bracken Cave Is an Incredible Natural Phenomenon with 20 Million Players," Texas Public Radio, July 16, 2024, *https://www.tpr.org/arts-culture/2024-07-16/batnado-at-bracken-cave-is-an-incredible-natural-phenomenon-with-twenty-million-players*.

57 **a thirty-pound newborn:** Author notes, Bracken Cave docent presentation, July 13, 2023.

57 **bats nursed indiscriminately:** Dr. Gary McCracken, professor emeritus, University of Tennessee, Knoxville, video call, July 9, 2024.

58 **feed their baby:** McCracken, video call, July 9, 2024.

58 **what one expert described as "a radiant heater":** Merlin Tuttle, founder, Merlin Tuttle's Bat Conservation, video call, December 18, 2023.

58 **150 tons of moths, beetles, and other insects:** Fran Hutchins, director, Bracken Cave Preserve, phone call, July 29, 2024.

59 **forty consecutive days:** Christopher Adams, "Austin's Record Streak of Consecutive 100° Days Finally Ends after 45 Days," KXAN, July 25, 2023, *https://www.kxan.com/weather/weather-blog/july-2023-100-degrees-streak/*.

59 **limited to fifteen minutes:** Jim Carlton, "Welcome to Summer Camp! You Can Play Outside for 15 Minutes," *The Wall Street Journal*, July 19, 2023, *https://www.wsj.com/articles/welcome-to-summer-camp-you-can-play-outside-for-15-minutes-ccf9ac28*.

59 **San Antonio City Councilman Ron Nirenberg:** Eileen Pace, "In Texas, the World's Biggest Bat Colony Is Saved from City Sprawl," NPR, November 1, 2014, *https://www.npr.org/2014/11/01/360481381/in-texas-a-massive-bat-nursery-is-saved-from-suburban-lights*.

60 **changing the cave's landscape:** JoAnna Wendel, "How Bat Breath and Guano Can Change the Shapes of Caves," *Eos*, November 9, 2015, *https://eos.org/articles/how-bat-breath-and-guano-can-change-the-shapes-of-caves*.

60 **largest mineral export:** Katie Armstrong, "Guano Gathering," *Texas Parks & Wildlife*, January 2007, *https://tpwmagazine.com/archive/2007/jan/scout3/*.

60 **can produce gunpowder:** Laurie E. Jasinki, "Confederate Bat Guano

Kiln, New Braunfels," Handbook of Texas Online, Texas State Historical Association, April 19, 2012, *https://www.tshaonline.org/handbook/entries/confederate-bat-guano-kiln-new-braunfels*.

60 **cleaned skeleton within hours:** Hutchins, phone call, July 29, 2024.

61 **[Cross section of Bracken Cave]:** Based on author's photo of public signage at Bracken Cave Preserve, July 13, 2023.

61 **2,500 pounds of guano:** Dr. Donald Frazier, director, the Texas Center, Schreiner University, email, August 16, 2024.

61 **can exceed six stories:** "Bracken and Its Bats: A Natural Wonder of the World," Bat Conservation International, June 29, 2017, *https://www.batcon.org/bracken-and-its-bats-a-natural-wonder-of-the-world/*.

61 **take off from the ground:** Bill Dowd, "How Do Bats Fly Compared to Birds?," Skedaddle Humane Wildlife Control, May 31, 2020, *https://www.skedaddlewildlife.com/location/whitby/blog/how-bats-fly-compared-to-birds/*.

61 **catch the wind like a paraglider:** Author notes, in-person docent presentation, Bracken Cave, July 13, 2023.

61 **99 miles per hour:** Julianna Photopoulos, "Speedy Bat Flies at 160km/h, Smashing Bird Speed Record," *New Scientist*, November 9, 2016, *https://www.newscientist.com/article/2112044-speedy-bat-flies-at-160kmh-smashing-bird-speed-record/*.

61 **242 miles per hour:** "Fastest Bird (Diving)," Guinness World Records, *https://www.guinnessworldrecords.com/world-records/70929-fastest-bird-diving*.

62 **124 feet wide and 117 feet tall:** "Beyond the Bats: Bracken Cave Preserve," Bat Conservation International, Facebook video, March 27, 2024, 51 min., 22 sec., timestamp: 4 min., 30 sec., *https://www.facebook.com/BatCon/videos/1533788843862607*.

62 **meaning Lady Liberty could:** "Statue Statistics" (heel to top of head), Statue of Liberty National Monument, National Park Service, *https://www.nps.gov/stli/learn/historyculture/statue-statistics.htm*.

62 **extending 650 feet:** Jeannie Ralston, "The Battle for Bracken Cave," *Nature Conservancy*, October/November 2015, 56, *https://www.nature.org/content/dam/tnc/nature/en/documents/Nature_Mag_2015_Bracken_Bats.pdf*.

63 **fifteen consecutive summers:** *Deep in the Heart*, directed by Ben Masters (2022, Deep in the Heart Film LLC), Prime Video.

63 **"appealing to potential immigrants":** Dr. Winifred Frick, chief scientist, Bat Conservation International, email, June 25, 2024.

63 **more than 1,400 species:** "13 Awesome Facts about Bats," US Department of the Interior, October 24, 2024, *https://www.doi.gov/blog/13-facts-about-bats*.

63 **One out of every five mammals:** Bill O'Brian, "The Coolest Mammals

on Earth," US Fish and Wildlife Service, n.d., *https://www.fws.gov/story/coolest-mammals-earth*.

63 **a mere six inches:** "Bumblebee Bat (Craseonycteris thonglongyai): Dimensions & Sizes," Dimensions.com, *https://www.dimensions.com/element/bumblebee-bat-craseonycteris-thonglongyai*.

63 **to nearly six feet:** "Roosting," Merlin Tuttle's Bat Conservation, 2006, *https://merlintuttle.smugmug.com/HighQuality/Roosting/i-zKgM3Mm*.

63 **collectively sense the nuances of the landscape:** Dr. Andrea Flack, group leader, Max Planck Institute of Animal Behavior, email, August 7, 2024.

64 **shared by scent:** Flack, email, August 7, 2024.

65 **"responding to the weather":** Shepard, email, August 16, 2024.

65 **More than one hundred thousand giant tortoises:** Rich Baxter, director, Indian Ocean Tortoise Alliance, email, August 9, 2024.

65 **2.5 million flamingos:** "Largest flamingo colony," Guinness World Records, *https://www.guinnessworldrecords.com/world-records/542210-largest-flamingo-colony*.

66 **five to thirty-five residents per acre:** Mary Beth Griggs and Daisy Hernandez, "10 Amazing Architects of the Animal Kingdom," *Popular Mechanics*, October 4, 2019, *https://www.popularmechanics.com/science/animals/g518/8-amazing-architects-of-the-animal-kingdom/?slide=1*.

66 **identify them from satellite imagery:** Eva Frederick, "Dog Town," *Texas Parks & Wildlife*, December 2018, *https://tpwmagazine.com/archive/2018/dec/LLL_prairiedogs/*.

66 **spanned 25,000 square miles:** Frederick, "Dog Town."

66 **forty times the size:** "Houston," Greater Houston Convention and Visitors Bureau, *https://www.houstontx.gov/planning/Demographics/demograph_docs/FactsFlier.pdf*.

66 **Dense animal cities are also more susceptible:** Tuttle, video call, December 18, 2023.

74 **animals' immediate extermination:** Rosie Ninesling, "The Year Austin Wanted to Exterminate the Bats," *Austin Monthly*, September 2019, *https://www.austinmonthly.com/the-year-austin-wanted-to-exterminate-the-bats/*.

74 **"dopes, villains, and terrible ideas":** KVUE Staff, "Texas Monthly's 'Bum Steer of the Year' Title Goes to . . . Lt. Gov. Dan Patrick," KVUE, December 11, 2023, *https://www.kvue.com/article/news/local/texas/texas-monthly-lt-gov-dan-patrick-bum-steer-of-the-year/269-c0e43290-e477-4fe9-bc74-feab77e98149*.

74 **TxDOT officials dispute:** Mark J. Bloschock, PE, retired, phone call, July 29, 2024.

75 **"more bang for your buck":** Bloschock, phone call, July 29, 2024.

75 **"Merlin Tuttle Day":** Duncan Hicks, operations manager, Merlin Tuttle's Bat Conservation, email, August 23, 2024.

78 **make final preparations:** Author notes, in-person visit to riverside educational exhibits in Austin, Texas, July 15, 2023.

78 **one such hub:** Author notes, in-person visit to riverside educational exhibits in Austin, Texas, July 15, 2023.

78 **causing Congress Avenue's population to swell:** Author notes, in-person visit to riverside educational exhibits in Austin, Texas, July 15, 2023.

79 **Austin Bat Refuge:** Lee Mackenzie and Dianne Odegard, secretary/treasurer and executive director, Austin Bat Refuge, video call, July 25, 2024.

79 **dialects or accents:** Rachael Lallensack, "Baby Bats Crowdsource Their Dialects from Colony Members," *Nature*, October 31, 2017, *https://doi.org/10.1038/nature.2017.22926*.

83 **100 tons of bats:** Hutchins, text message, August 24, 2024.

84 **[Map of bat emergences]:** Doppler data: NOAA Multi-Radar/Multi-Sensor (MRMS) Merged Reflectivity Composite, *https://tinyurl.com/y8w6aht6*. Locations and timestamps: Austin Bat Refuge, *https://austinbatrefuge.org/radar/*.

3. E PLURIBUS UNUM: NATIONS

87 **volcanic island of Madeira:** Juan José Gonçalves Silva, curator of botany, and Ysabel Margarita Amaro Gonçalves, curator of entomology, Museu de História Natural do Funchal, email, August 6, 2024.

88 **Isola della Lolegname:** Author notes, in-person visit to Madeira Story Centre, Funchal, Portugal, September 24, 2023.

88 **Eucalyptus and acacia trees:** Silva and Gonçalves, email, August 6, 2024.

88 **trade ship from Argentina:** James K. Wetterer et al., "Worldwide Spread of the Argentine Ant, Linepithema humile (Hymenoptera: Formicidae)," *Myrmecological News* 12 (October 1, 2009): 187–94, *https://www.researchgate.net/publication/228654113_Worldwide_spread_of_the_Argentine_ant_Linepithema_humile_Hymenoptera_Formicidae*.

89 **Great Chain of Being:** S. Snyder, "The Great Chain of Being," Grand View University, archived, retrieved January 2017, *https://web.archive.org/web/20170728101006/http://faculty.grandview.edu/ssnyder/121/121%20great%20chain.htm*.

89 **unpredictable sea monsters:** Chet Van Duzer, *Sea Monsters on Medieval and Renaissance Maps* (British Library, 2013), 71–76.

89 **the term *civilization*:** Cynthia Stokes Brown, "What Is a Civilization, Anyway?," World History Connected, October 2009, *https://worldhistoryconnected.press.uillinois.edu/6.3/brown.html*.

90 **[Waldseemüller map (detail)]:** Martin Waldseemüller (cartographer) and Johann Schöner, *Carta marina navigatoria Portvgallen navigationes, atqve tocius cogniti orbis terre marisqve* . . . (Strasbourg, France [?]: publisher not

identified, 1516). From Library of Congress, Geography and Map Division, *https://www.loc.gov/item/2016586433/*.

90 **deemed a "backward" society:** National Geographic Society, "Civilizations," *National Geographic*, *https://education.nationalgeographic.org/resource/civilizations/*.

90 **lack of a written alphabet and non-Christian belief system:** Terence N. D'Altroy, founding director, Columbia Center for Archaeology, email, January 14, 2025.

90 **nomads like Bedouins and Kazakhs:** Dr. Steve Sabol, professor, Department of History, University of North Carolina at Charlotte, Facebook direct message, December 12, 2024.

90 **city of Great Zimbabwe:** Zeb Larson, "The Ancient Remains of Great Zimbabwe," BBC, September 26, 2022, *https://www.bbc.com/travel/article/20220925-the-ancient-remains-of-great-zimbabwe*.

90 **expanded the definition:** National Geographic Society, "Civilizations."

91 **they share land:** Cecilia Keating, "Indigenous Geographies Overlap in This Colorful Online Map," Atlas Obscura, July 24, 2018, *https://www.atlasobscura.com/articles/native-land-map-of-indigenous-territories*.

91 **fewer than thirty people:** Author notes, in-person visit to Monument Valley, Navajo Nation, March 31, 2024.

92 **longer than a football field:** "Funchalense 5: General Cargo," vessel characteristics, MarineTraffic, *https://www.marinetraffic.com/en/ais/details/ships/shipid:299922/mmsi:255804250/imo:9388390/vessel:FUNCHALENSE_5*.

92 **hundreds of identical shipping containers:** Carolina Nunes, commercial assistant, Grupo Sousa—GS Lines, email, July 29, 2024.

92 **case around 1858:** Wetterer et al., "Worldwide Spread of the Argentine Ant."

92 **spans 3,600 miles:** Susan Fineman, "Largest Supercolony of Ants Found in Spain," *The Washington Post*, April 20, 2002, *https://www.washingtonpost.com/archive/local/2002/04/21/largest-supercolony-of-ants-found-in-spain/4ea2bb93-69ee-49a7-a615-d82c79c4df5e/*.

92 **billions of individuals:** Dr. Giacomo Santini, associate professor of ecology and head of the research group, Università degli Studi di Firenze, email, July 8, 2024.

92 **"largest cooperative unit ever recorded":** Tatiana Giraud, Jes S. Pedersen, and Laurent Keller, "Evolution of Supercolonies: The Argentine Ants of Southern Europe," *Proceedings of the National Academy of Sciences* 99, no. 9 (April 30, 2002): 6075–79, *https://doi.org/10.1073/pnas.092694199*.

93 **"scourge in the Mediterranean":** Ana Isabel Queiroz, "Scourge in the Mediterranean. Policies Against the Argentine Ant in the First Half of the 20th Century," in *Histories of Bioinvasions in the Mediterranean*, ed.

Ana Isabel Queiroz and Simon Pooley (Springer International, 2018), 87–103, *https://doi.org/10.1007/978-3-319-74986-0_4*.

93 **"unique among injurious insects":** Wilmon Newell and T. C. Barber, *The Argentine Ant* (US Department of Agriculture, Bureau of Entomology, 1913), 3.

93 **"household pest of the first rank":** Newell and Barber, *The Argentine Ant*, 3.

93 **"been known to eat a baby":** Ana Isabel Queiroz and Daniel Alves, "People, Transports and the Spread of the Argentine Ant in Europe, from c. 1850 to Present," *CEM/Cultura, Espaço & Memória* 7 (June 1, 2016): 37–62, *https://www.researchgate.net/publication/312198027_People_transports_and_the_spread_of_the_Argentine_ant_in_Europe_from_c_1850_to_present*.

94 **[Map of Argentine ant supercolonies]:** Data courtesy of Benoit Guénard and also appears in Elena Angulo et al., "The Argentine Ant, *Linepithema humile*: Natural History, Ecology and Impact of a Successful Invader," *Entomologia Generalis* 44, no.1 (February 23, 2024): 41–61, *https://doi.org/10.1127/entomologia/2023/2187*. Additional nest locations and guidance courtesy of Drs. Xavier Espadaler, Giacomo Santini, Elena Angulo, and Olivier Blight. Where specific ant populations have not been genetically tested, assumptions were made by the researchers about which supercolonies the ants likely belong to.

94 **combat the unsuspecting native species:** Santini, email, July 8, 2024.

94 **wine, sugarcane, or even South American tourists:** Dr. Silvia Abril Melendez, Department of Environmental Sciences, University of Girona, email, September 1, 2024.

96 **"advantageous to align with":** Dr. Chase L. Núñez, postdoctoral researcher, Department for the Ecology of Animal Societies, Max Planck Institute of Animal Behavior, email and video call, December 6, 2024.

96 **animal nations start to take shape:** Núñez, video call, December 6, 2024.

97 **never exceeded 553:** Ray Thompson, volunteer, Death Valley National Park, who consulted with the national park's fisheries biologist Jeffrey Goldstein and supervisory biologist Kevin P. Wilson, PhD, phone call, July 21, 2024.

97 **"more ancient than Sanskrit":** Ross Andersen, "How First Contact with Whale Civilization Could Unfold," *The Atlantic*, February 24, 2024, *https://www.theatlantic.com/science/archive/2024/02/talking-whales-project-ceti/677549/*.

98 **"little puddles of shade":** Ari Daniel, "These Monkeys Were 'Notoriously Competitive' until Hurricane Maria Wrecked Their Home," NPR, July 10, 2024, *https://www.npr.org/2024/07/09/g-s1-6734/rhesus-macaques-hurricane-maria-changed-monkey-society*.

98 **caterpillars, who consume water:** "Do Caterpillars Drink Water?," The Children's Butterfly Site, *https://www.kidsbutterfly.org/faq/behavior/8*.

98 **drink and carry back to their nests:** Alberto Masoni, postdoctoral researcher, Università degli Studi di Firenze, email, December 19, 2024.

99 **maintain millions of nests:** Fineman, "Largest Supercolony of Ants Found in Spain."

100 **waging mass executions:** Sílvia Abril and Crisanto Gómez, "Factors Triggering Queen Executions in the Argentine Ant," *Scientific Reports* 9, no. 1 (July 18, 2019): 10427, *https://doi.org/10.1038/s41598-019-46972-5*.

100 **annual coup d'etat:** Melendez, video call, May 12, 2023.

101 **"without assistance from the parent nest":** Newell and Barber, *The Argentine Ant*, 19.

101 **derogatory to Romani people:** Sabrina Imbler, "This Moth's Name Is a Slur. Scientists Won't Use It Anymore," *The New York Times*, July 9, 2021, *https://www.nytimes.com/2021/07/09/science/gypsy-moth-romanientomological-society.html*.

101 **European researchers don't appear to share this concern:** Dr. Elena Angulo, Estación Biológica de Doñana, WhatsApp message, January 5, 2025.

103 **restricting highway access:** Mariana Pereyra and Alejandro G. Farji-Brener, "Traffic Restrictions for Heavy Vehicles: Leaf-Cutting Ants Avoid Extra-Large Loads When the Foraging Flow Is High," *Behavioural Processes* 170 (January 1, 2020): 104014, *https://doi.org/10.1016/j.beproc.2019.104014*.

103 **additional lanes will be formed:** Tanya Latty et al., "Argentine Ants (*Linepithema humile*) Use Adaptable Transportation Networks to Track Changes in Resource Quality," *Journal of Experimental Biology* 220, no. 4 (February 15, 2017): 686–94, *https://doi.org/10.1242/jeb.144238*.

103 **"I am not so important":** Prof. Xavier Espadaler, Universitat Autònoma de Barcelona, email, July 8, 2024.

104 **"extremely high" aggression:** Giraud, Pedersen, and Keller, "Evolution of Supercolonies."

104 **viewed not as brothers and sisters:** Dr. Tatiana Giraud, director of research, Centre National de la Recherche Scientifique, email, February 9, 2025.

104 **98 percent of cases:** Giraud, Pedersen, and Keller, "Evolution of Supercolonies."

105 **as "trench warfare":** "UCSD Study Says Argentine Ants Very Territorial," *The San Diego Union-Tribune*, December 10, 2006, *https://www.sandiegouniontribune.com/2006/12/10/ucsd-study-says-argentine-ants-very-territorial/*.

105 **third Argentine ant supercolony:** O. Blight et al., "A New Colony Structure of the Invasive Argentine Ant (*Linepithema humile*) in Southern Europe," *Biological Invasions* 12, no. 6 (June 1, 2010): 1491–97, *https://doi.org/10.1007/s10530-009-9561-x*.

106 **origin story of the Corsicans:** Dr. Olivier Blight, senior lecturer, Mediterranean Institute of Biodiversity and Ecology, email, July 16, 2024.

107 **"What Is a Species, Anyway?":** Carl Zimmer, "What Is a Species, Anyway?," *The New York Times*, February 19, 2024, *https://www.nytimes.com/2024/02/19/science/what-is-a-species.html*.

107 **splinter orca whales:** Emily Anthes, "All Orcas Are Classified as a Single Species. Should They Be?," *The New York Times*, March 26, 2024, *https://www.nytimes.com/2024/03/26/science/orcas-species-killer-whales.html*.

108 **notice such differences long before we catch on:** Dr. Deborah Giles, science and research director, Wild Orca, phone call, September 8, 2024.

108 **described as a "peace zone":** Laurence Berville et al., "A Peaceful Zone Bordering Two Argentine Ant (*Linepithema humile*) Supercolonies," *Chemoecology* 23, no. 4 (December 1, 2013): 213–18, *https://doi.org/10.1007/s00049-013-0135-0*.

112 **that E. O. Wilson noted:** Edward O. Wilson, *In Search of Nature* (Island Press, 1996), 47–48.

115 **known as Lanchester's laws:** Masoni, email, March 12, 2025.

115 **ten workers and one queen:** Sara Castro-Cobo et al., "Long-Term Spread of Argentine Ant (Hymenoptera: Formicidae) European Supercolonies on Three Mediterranean Islands," *Myrmecological News* 31 (2021): 185–200, *https://doi.org/10.25849/myrmecol.news_031:185*.

116 **no continent but Antarctica:** Chris Scocco, "Argentine Ants Come Inside for Warmth, Food and Shelter," College of Agricultural and Environmental Sciences Newswire, University of Georgia, January 28, 2010, *https://newswire.caes.uga.edu/story/3696/marching-ants.html*.

116 **Intercontinental Union of Argentine Ants:** Eiriki Sunamura et al., "Intercontinental Union of Argentine Ants: Behavioral Relationships among Introduced Populations in Europe, North America, and Asia," *Insectes Sociaux* 56, no. 2 (July 1, 2009): 143–47, *https://doi.org/10.1007/s00040-009-0001-9*.

116 **repel further intrusions:** Olivier Blight et al., "A Native Ant Armed to Limit the Spread of the Argentine Ant," *Biological Invasions* 12, no. 11 (November 1, 2010): 3785–93, *https://doi.org/10.1007/s10530-010-9770-3*.

117 **unclear to him who's fighting who:** Blight, email, July 19, 2024.

117 **enabled hunter-gatherers to:** Paul Ratner, "How Engineering by Beavers Helped the Survival of Ancient Humans," Interesting Engineering, November 17, 2023, *https://interestingengineering.com/science/engineering beavers-survival-ancient-humans*.

118 **at 20 quadrillion:** Amarachi Orie, "Scientists Have Estimated How Many Ants There Are on Earth. Clue: It's a Lot," CNN, September 20, 2022, *https://www.cnn.com/2022/09/20/world/ants-twenty-quadrillion-earth-intl-scli-scn/index.html*.

118 **amputees to return to duty:** Ashley Strickland, "Carpenter Ants Amputate the Legs of Their Nestmates to Save Their Lives, Study Says," CNN, July 3, 2024, *https://www.cnn.com/2024/07/03/science/ant-leg-amputations-scn/index.html*.

118 **red fire ant:** Katie Hunt, "'We Knew This Day Would Come': One of World's Most Invasive Species Reaches Europe," CNN, September 11, 2023, *https://www.cnn.com/2023/09/11/europe/invasive-red-fire-ant-europe-italy-scn/index.html*.

118 **95 percent of Europe:** Raymond Zhong, "Three-Quarters of Earth's Land Got Drier in Recent Decades, U.N. Says," *The New York Times*, December 9, 2024, *https://www.nytimes.com/2024/12/09/climate/global-desertification.html*.

119 **"cradle of civilization":** Alissa J. Rubin, "A Climate Warning from the Cradle of Civilization," *The New York Times*, July 29, 2023, *https://www.nytimes.com/2023/07/29/world/middleeast/iraq-water-crisis-desertification.html*.

119 **desperate migrant groups:** Rebecca Geldard, "Here's How Extreme Weather Is Affecting Animal Migration," World Economic Forum, October 5, 2023, *https://www.weforum.org/stories/2023/10/climate-crisis-impacting-animal-migration/*.

119 **"more diversity in a single tree":** Dr. Neil D. Tsutsui, professor and Michelbacher Chair of Systematic Entomology, University of California, Berkeley, email, July 1, 2024.

119 **native ant communities swiftly returned:** Meghan Cooling et al., "The Widespread Collapse of an Invasive Species: Argentine Ants (*Linepithema humile*) in New Zealand," *Biology Letters* 8, no. 3 (November 30, 2011): 430–33, *https://doi.org/10.1098/rsbl.2011.1014*.

4. BELOW THE SURFACE: LAYERS

122 **about a foot below humanity's feet since the dawn of our species:** Kyle Finn, postdoctoral fellow, Department of Zoology and Entomology, University of Pretoria, email, August 6, 2024.

122 **northern Zambia and in parts of neighboring Angola and the Democratic Republic of the Congo:** Paul A. A. G. Daele et al., "A New Species of African Mole-Rat (*Fukomys*, Bathyergidae, Rodentia) from the Zaire-Zambezi Watershed," *Zootaxa* 3636 (April 3, 2013): 171–89, figure 1, *https://doi.org/10.11646/zootaxa.3636.1.7*.

122 **staggering in scope, even to them:** Dr. Radim Šumbera, professor, Department of Zoology, University of South Bohemia, email, July 15, 2024.

122 **more than a month of digging:** Šumbera, email, July 15, 2024.

122 **largest of its kind:** Šumbera, email, July 15, 2024.

123 **possibly even decades:** Finn, email, July 1, 2024.

123 **four or five bouts of activity:** Finn, email, July 1, 2024.

124 **[Map of Ndola tunnel network]:** Radim Šumbera et al., "Burrow Architecture, Family Composition and Habitat Characteristics of the Largest Social African Mole-Rat: The Giant Mole-Rat Constructs Really Giant Burrow Systems," *Acta Theriologica* 57, no. 2 (April 1, 2012): 121–30, *https://*

doi.org/10.1007/s13364-011-0059-4. Additional notations and data provided by Šumbera and Vladimír Mazoch on January 4, 2024.

125 **metabolize into water:** Finn, email, August 6, 2024.

126 **call an "overthrow":** Finn, email, July 1, 2024.

126 **groundhogs spend the majority:** Dr. Greg Florant, emeritus professor of biology, Colorado State University, email, December 3, 2024.

127 **more than half of all life:** Phoebe Weston, "More Than Half of Earth's Species Live in the Soil, Study Finds," *The Guardian*, August 7, 2023, *https://www.theguardian.com/environment/2023/aug/07/more-than-half-of-earths-species-live-in-the-soil-study-finds-aoe*.

127 **by an insectivorous mole:** Vladimir Dinets, research assistant professor, University of Tennessee, Knoxville, email, December 3, 2024.

127 **called the "subnivean" zone:** Bethany Heft, "The Subnivean Layer—Ever Heard of It?," Trees for Tomorrow, *https://treesfortomorrow.com/Blog/Survival-In-the-Snow/*.

127 **Similar stratification takes place up in the treetops:** Núñez, email, December 17, 2024.

128 **distinct populations of warblers:** Rana Ahmed, "Interspecific Competition," Study.com, November 21, 2023, *https://study.com/learn/lesson/interspecific-competition-competitive-exclusion-niche-differentiation.html*.

128 **clan of the Kombai people:** Oliver Steeds, "The Kombai," *Avaunt Magazine*, no. 1 (Spring/Summer 2015), *https://avauntmagazine.com/the-kombai/*.

128 **majority of the town's 1,500 residents:** "2021 Coober Pedy, Census All Persons QuickStats," Australian Bureau of Statistics, *https://abs.gov.au/census/find-census-data/quickstats/2021/LGA41330*.

128 **approach of moving underground:** Zaria Gorvett, "The Australian Town Where People Live Underground," BBC, August 3, 2023, *https://www.bbc.com/future/article/20230803-the-town-where-people-live-underground*.

128 **churches, an art gallery, and a pub:** Catherine Bell, executive assistant, District Council of Coober Pedy, email, January 16, 2025.

128 **tens of thousands of ancient peoples:** Geena Truman, "Turkey's Underground City of 20,000 People," BBC, August 11, 2022, *https://www.bbc.com/travel/article/20220810-derinkuyu-turkeys-underground-city-of-20000-people*.

129 **limestone, coal, manganese, and gold:** "Zambia 200TPH Limestone Crushing Plant," SBM Industrial Technology Group, SBM, *https://www.sbmchina.com/case/aggregate/1146.html*; "China's Mining Footprint in Africa," in *China in Africa*, AGE (African Growing Enterprises) File, Institute of Developing Economies–Japanese External Trade Organization, 2009, *https://www.ide.go.jp/English/Data/Africa_file/Manualreport/cia_08.html*; William Gu, "China Mineral Resources Zambian Kashi Copper and Gold Mine Mining Cooperation Project Completed and Put into Production," *SMM Synthesis*, December 27, 2021, *https://news*

.metal.com/newscontent/101706566/china-mineral-resources-zambian-kashi-copper-and-gold-mine-mining-cooperation-project-completed-and-put-into-production.

129 **biologist Francis E. Putz:** Sofia Quaglia, "Animals Developed Agriculture Before Us. These Gophers May 'Farm' as Well," *National Geographic*, July 11, 2022, *https://www.nationalgeographic.com/animals/article/pocket-gophers-farm-roots-study-suggests*.

131 **magnetic field as a compass:** Finn, email, December 2, 2024.

131 **during the Pleistocene Epoch:** Šumbera, email, August 19, 2024.

131 **roughly 1.3 million years:** Dr. Rachel Narducci, vertebrate paleontology collections manager, Florida Museum of Natural History, email, August 20, 2024.

131 **little over 10,000 years ago:** National Geographic Society, "The Development of Agriculture," *National Geographic*, November 2024, *https://education.nationalgeographic.org/resource/development-agriculture/*.

132 **Florida Farm Bureau reiterated:** Oliver Whang, "The Question You Didn't Know Needed Answering: Are Gophers Farmers?," *The New York Times*, July 11, 2022, *https://www.nytimes.com/2022/07/11/science/gophers-farmers-roots.html* (emphasis added).

132 **"because it's not really settled":** Francis E. Putz, Distinguished Professor, Department of Biology, University of Florida, email, June 30, 2024.

133 **including leafcutter ants:** "Leafcutter Ant," Rainforest Alliance, December 10, 2022, *https://www.rainforest-alliance.org/species/leafcutter-ant/*.

133 **and some termites:** Sarah Kaplan, "Termites Figured Out Farming 25 Million Years Before People Did," *The Washington Post*, June 23, 2016, *https://www.washingtonpost.com/news/speaking-of-science/wp/2016/06/23/termites-figured-out-farming-25-million-years-before-people-did/*.

133 **some snails farm too:** Brian R. Silliman and Steven Y. Newell, "Fungal Farming in a Snail," *Proceedings of the National Academy of Sciences* 100, no. 26 (December 23, 2003): 15643–48, *https://doi.org/10.1073/pnas.2535227100*.

133 **known as *Polysiphonia* sp. 1:** Dr. Hiroki Hata, professor, Ehime University, email, July 1, 2024.

134 **[Damselfish image]:** Photo published with permission, courtesy of Yuhei Tsuchiya, Itoman Diving Service, Okinawa, Japan, December 7, 2023.

134 **damselfish have in effect "domesticated" the shrimp**: Rohan M. Brooker et al., "Domestication via the Commensal Pathway in a Fish-Invertebrate Mutualism," *Nature Communications* 11, no. 1 (December 7, 2020): 6253, *https://doi.org/10.1038/s41467-020-19958-5*.

134 **10 percent of ocean life:** Rebecca Carballo, "Scientists Discover 100 New Marine Species in New Zealand," *The New York Times*, March 10, 2024, *https://www.nytimes.com/2024/03/10/science/new-species-sea-discovery.html*.

135 **Since the Stone Age:** C. S. Henshilwood, "Identifying the Collector: Evidence for Human Processing of the Cape Dune Mole-Rat, *Bathyergus suillus*, from Blombos Cave, Southern Cape, South Africa," *Journal of Archaeological Science* 24, no. 7 (July 1, 1997): 659–62, *https://doi.org/10.1006/jasc.1996.0148*.

136 **kind of life raft:** Finn, email, August 6, 2024.

136 **four miles of tunnels:** Michael Marshall, "Zoologger: The Longest Tunnels Dug by a Mammal," *New Scientist*, August 10, 2012, *https://www.newscientist.com/article/dn22164-zoologger-the-longest-tunnels-dug-by-a-mammal/*.

138 **almost doubled in the past twenty years:** "Zambia Population," Worldometer, 2025, *https://www.worldometers.info/world-population/zambia-population/*.

138 **equivalent of a New York City every month:** "Why the Built Environment?," Architecture 2030, *https://www.architecture2030.org/why-the-built-environment/*.

139 **five times the size of Central Park:** "Park History," Central Park Conservancy, *https://www.centralparknyc.org/park-history*.

139 **fewer than 700:** Paul Golson, Lusaka-based conservationist, Facebook direct message, December 15, 2024.

140 **new kinds of mole-rats:** Daele et al., "A New Species of African Mole-Rat."

141 **ground-penetrating radar (GPR):** Timothy Saey et al., "Reconstructing Mole Tunnels Using Frequency-Domain Ground Penetrating Radar," *Applied Soil Ecology* 80 (August 1, 2014): 77–83, *https://doi.org/10.1016/j.apsoil.2014.03.019*.

141 **two inches wide:** Niklas Allroggen et al., "High-Resolution Imaging and Monitoring of Animal Tunnels Using 3D Ground-Penetrating Radar," *Near Surface Geophysics* 17, no. 3 (June 1, 2019): 291–98, *https://doi.org/10.1002/nsg.12039*.

141 **survey the same spot over and over again:** Kirstin Deuss, pedologist, Manaaki Whenua–Landcare Research, email, June 27, 2024.

5. SAFE PASSAGE: TRANSIT

146 **antlers 12 feet across:** Amanda Ellis, "This Giant Deer Species Went Extinct Wielding 12-Ft Antlers," Roaring Earth, March 20, 2024, *https://roaring.earth/giant-deer*.

146 **considered the saigas to be their guides:** Saiga Alliance, "Saigachy Documentary—English," YouTube, posted January 12, 2016, 27 min., 3 sec., timestamp: 7 min., 50 sec., *https://www.youtube.com/watch?app=desktop&v=FwcgAGN2CaM*.

147 **among the first species to be monitored from space:** Dr. Martin Wikelski, director, Department of Migration, Max Planck Institute of Animal Behavior, email, July 23, 2024.

147 **AI researchers in Russia:** Viatcheslav Rozhnov, "Remote and Artificial Intelligence Methods to Estimate the Saiga Population in North-Western Caspian Sea Area and How Animals Use the Area," *Saiga News*, no. 29 (2023): 26–29.

148 **Early Iron Age:** Dr. Dmitriy Voyakin, Goodwill Ambassador of Kazakhstan, coordinator to the University College London Central Asian Archaeological Landscapes project, email, June 28, 2024.

148 **Turkic word *qaz*:** Voyakin, email, August 9, 2024.

149 **wrestling and tug-of-war:** Voyakin, email, August 9, 2024.

149 **pave the way:** Michael D. Frachetti et al., "Nomadic Ecology Shaped the Highland Geography of Asia's Silk Roads," *Nature* 543, no. 7644 (March 1, 2017): 193–98, *https://doi.org/10.1038/nature21696*.

149 **also a key intersection:** "Kazakhstan," Silk Roads Programme," UNESCO, *https://en.unesco.org/silkroad/countries-alongside-silk-road-routes/kazakhstan*.

149 **seventh century BCE:** G. Bazarbayeva and G. Jumabekova, "The Image of Antelope (Saiga) in the Early Iron Age Art of Kazakhstan," *Proceedings in Archaeology and History of Ancient and Medieval Black Sea Region* no. 15 (n.d.): 893–907, *https://doi.org/10.53737/3051.2023.74.84.037*.

149 **the 2,000-tenge bill:** Ramilya Sazazova and Michelle Blau, "Save the Saiga Antelope," *FrontLines*, U.S. Agency for International Development, November/December 2013, archived, *https://web.archive.org/web/20240615033918/https://2012-2017.usaid.gov/news-information/frontlines/depleting-resources/save-saiga-antelope*.

150 **force them to farm:** Christopher Robbins, *Apples Are from Kazakhstan: The Land That Disappeared* (W. W. Norton, 2008), 24–25.

150 **burned brightly for millennia:** Julio Bendezu Sarmiento, "The First Nomads in Central Asia's Steppes (Kazakhstan)," in *Nomad Lives: From Prehistoric Times to the Present Day*, ed. Aline Averbough, Nejma Goutas, and Sophie Méry (Publications scientifiques du Muséum, 2022), *https://books.openedition.org/mnhn/11725?lang=en*.

150 **much of the steppe soil was unsuitable for growing crops:** Robbins, *Apples Are from Kazakhstan*, 24.

150 **Werner Herzog once remarked:** Paul Theroux, *The Tao of Travel: Enlightenments from Lives on the Road* (Houghton Mifflin Harcourt, 2011), 137.

150 **stationary lives in urban centers:** "Urban population (% of total population)—Kazakhstan," World Bank Group, *https://data.worldbank.org/indicator/SP.URB.TOTL.IN.ZS?locations=KZ*.

152 **[Map of saiga movements]:** Tracking data courtesy of Altyn Dala Conservation Initiative research fellow Steffen Zuther, based on work con-

ducted as part of the Altyn Dala Conservation Initiative, in cooperation with the Committee of Forestry and Wildlife of the Ministry of Ecology and Natural Resources of the Republic of Kazakhstan, supported by the international conservation organizations Fauna & Flora, Frankfurt Zoological Society, and the Royal Society for the Protection of Birds.

152 **straddle the known ranges of local wolves:** Dr. Jose A. Hernandez-Blanco, senior scientific researcher, A. N. Severtsov Institute of Ecology and Evolution, Russian Academy of Sciences, email, August 6, 2024.

153 **"erasure of a language":** Ben Goldfarb, *Crossings: How Road Ecology Is Shaping the Future of Our Planet* (W. W. Norton, 2023), 40.

153 **active participation and observation:** Albert Salemgareyev, lead specialist, Association for the Conservation of Biodiversity of Kazakhstan, video call, August 11, 2024.

153 **Frederick Burnaby feared this wind:** Robbins, *Apples Are from Kazakhstan*, 132.

155 **Forest elephants have been shown:** Allard Blom, vice president, global integrated programs, African forests, World Wildlife Fund (via Irene Serrano, US media and external affairs specialist), email, July 23, 2024.

155 **compress the soil and create canals:** Mark Mancini, "Eco Engineers: 5 Animals That Can Reshape Earth's Waterways," HowStuffWorks, *https://animals.howstuffworks.com/animal-facts/5-animals-that-can-reshape-waterways.htm#pt4*.

155 **Arctic caribou also create:** Dr. Cyrus Harris, vice chair, Western Arctic Caribou Herd Working Group, email, July 13, 2024.

155 **"surfing the green wave":** Jason G. Goldman, "Sheep Teach Each Other How to Migrate Long Distances," *National Geographic*, September 6, 2018, *https://www.nationalgeographic.com/animals/article/bighorn-sheep-migration-culture-teaching-learning-news*.

155 **"expunge generations of knowledge":** Brett R. Jesmer et al., "Is Ungulate Migration Culturally Transmitted? Evidence of Social Learning from Translocated Animals," *Science* 361, no. 6406 (September 7, 2018): 1023–25, *https://doi.org/10.1126/science.aat0985*.

156 **Atlas of Ungulate Migration:** Zuther, email, December 10, 2024.

156 **famed Wilderness Road:** Dr. Lawrence J. Fleenor Jr., historian, Daniel Boone Wilderness Trail Association, email, July 10, 2024.

156 **paths of the Grand Canyon:** Dr. Paul Hirt, professor emeritus, Arizona State University, and director, Nature, Culture, and History at the Grand Canyon, email, August 21, 2024.

156 **kangaroos often follow:** Dr. Lynne Kelly, expert on Australian Aboriginal oral cultures, adjunct research fellow, School of Humanities and Social Sciences, La Trobe University, email, December 19, 2024.

156 **known as "auto trails":** Daniel P. Faigin, "Trails and Roads," California Highways, last modified August 13, 2024, *https://www.cahighways.org/trailroads.html*.

156 **pigeons and bees:** David Grimm, "Birds in the Fast Lane," *Science*, July 26, 2004, *https://www.science.org/content/article/birds-fast-lane*; Ian Randall, "Taking the B-Road! Bumblebees Use Paths and Lanes Created by Humans to Navigate Their Way Around, Study Finds," *The Daily Mail*, August 13, 2021, *https://www.dailymail.co.uk/sciencetech/article-9890443/Nature-Bumblebees-use-paths-lanes-created-humans-navigate-way-study-finds.html*.

156 **imprinting a map:** Ingrid Newkirk and Gene Stone, *Animalkind: Remarkable Discoveries about Animals and the Remarkable Ways We Can Be Kind to Them* (Simon & Schuster, 2020), 28–29.

156 **Baboon troops:** Ariana Strandburg-Peshkin et al., "Habitat and Social Factors Shape Individual Decisions and Emergent Group Structure during Baboon Collective Movement," ed. Catherine Carr, *eLife* 6 (January 31, 2017): e19505, *https://doi.org/10.7554/eLife.19505*.

156 **use "navigation memory":** Mischa Dijkstra, "Bees Follow Linear Landmarks to Find Their Way Home, Just like the First Pilots," Frontiers Science Communication, March 5, 2023, *https://www.frontiersin.org/news/2023/03/06/frontiers-behavioral-neuroscience-honeybees-navigate-by-linear-landscape-elements*.

157 **White-tailed deer:** Dr. James C. Kroll, director, Institute for White-Tailed Deer Management and Research, email, April 20, 2025.

157 **"its foremost trailblazers":** Robert Moor, *On Trails: An Exploration* (Simon & Schuster, 2016), 20.

157 **Botswana's Okavango Delta:** Moor, *On Trails*, 97–99.

158 **capturing sprinting saigas by the tens of thousands:** Saiga Alliance, "Saigachy Documentary—English," timestamp: 4:30–7:03.

159 **mass-exterminating whole saiga caravans:** "Ustyurt: Natural Landscape and Aran Hunting Traps," UNESCO World Heritage Convention, *https://whc.unesco.org/en/tentativelists/6571*.

159 **[Diagram of an aran]:** Inspired by "Saigachy Documentary—English."

160 **thousands of years ago:** "Ustyurt: Natural Landscape and Aran Hunting Traps," UNESCO World Heritage Convention.

161 **slightest scientific scrutiny:** Lalita Gomez and Kanitha Krishnasamy, "A Rapid Assessment of the Trade in Saiga Antelope Horn in Peninsular Malaysia," *Traffic Bulletin* 31, no. 1 (2019): 35–38, *https://www.traffic.org/site/assets/files/12036/saiga-horn-malaysia.pdf*.

161 **"critically endangered" status:** "Saiga Antelope: A Conservation Success Story," US Fish and Wildlife Service, *https://www.fws.gov/story/saiga-antelope-conservation-success-story*. (Cited source was adapted with permission from the Saiga Conservation Alliance by SCCF Program Officer Tatiana Hendrix.)

161 **one local ranger:** "Tragic News: Ranger Killed," Saiga Conservation Alliance, July 24, 2019, *https://saiga-conservation.org/2019/07/24/tragic-news-ranger-killed/*.

162 **more than a million today:** Salemgareyev, email, August 11, 2024.

163 **wiping out 200,000 saigas:** Aruzhan Ualikhanova, "Saiga Crisis: Farmers Seek Regulations amid Population Surge in Kazakhstan," *The Astana Times*, October 8, 2023, *https://astanatimes.com/2023/10/saiga-crisis-farmers-seek-regulations-amid-population-surge-in-kazakhstan/*.

169 **metallurgy to pottery:** Author notes, in-person visit to exhibit at the Atyrau Regional Museum, November 26, 2023.

171 **Prince Peter Kropotkin:** Dr. Lee Alan Dugatkin, animal behaviorist, evolutionary biologist, and historian of science, Department of Biology, University of Louisville, email, January 14, 2025.

171 **"do not let man approach their herds":** Peter Kropotkin, *Mutual Aid: A Factor of Evolution* (Freedom Press, 1987), ch. 1, *https://www.marxists.org/reference/archive/kropotkin-peter/1902/mutual-aid/ch01.htm*.

171 **beyond mere instinct:** J. P. Harpignies, "The Prince of Mutual Aid: Homage to Pyotr Alexeyevich Kropotkin," Bioneers, June 2, 2020, *https://bioneers.org/mutual-aid-homage-pyotr-alexeyevich-kropotkin/*.

172 **[Map of saiga ranges]:** Range data: Zuther. Silk road routes: Harvard WorldMap, accessed September 4, 2021, *https://worldmap.harvard.edu/*.

175 **the fourth-largest lake:** Victoria Steckhan, "The Disappearance of Earth's 4th Largest Lake," European Wilderness Society, *https://wilderness-society.org/the-disappearance-of-earths-fourth-largest-lake/*.

175 **shrank to a tenth:** Ankit Panda, "How the Soviet Union Created Central Asia's Worst Environmental Disaster," *The Diplomat*, October 3, 2014, *https://thediplomat.com/2014/10/how-the-soviet-union-created-central-asias-worst-environmental-disaster/*.

175 **split into four small ones:** "Studying the Aral Sea's Transformation Journey," EOS Data Analytics, last modified September 7, 2024, *https://eos.com/blog/the-aral-seas-rebirth-hope-blooms/*.

175 **largely abandoned the region:** Jacob Dreyer, "A Giant Inland Sea Is Now a Desert, and a Warning for Humanity," *The New York Times*, November 28, 2023, *https://www.nytimes.com/2023/11/28/opinion/climate-uzbekistan-water-aral.html*.

176 **"100 miles from its original shoreline":** Robbins, *Apples Are from Kazakhstan*, 121.

176 **revitalize and preserve:** Levi Bridges and Leia Larsen, "Kazakhstan's Drying Aral Sea Carries a Message for Those Worried about the Great Salt Lake," KUER, January 15, 2024, *https://www.kuer.org/health-science-environment/2024-01-15/kazakhstans-drying-aral-sea-carries-a-message-for-those-worried-about-the-great-salt-lake*.

176 **The summer of 2014:** "The Aral Sea Loses Its Eastern Lobe," Earth Observatory, September 26, 2014, *https://earthobservatory.nasa.gov/images/84437/the-aral-sea-loses-its-eastern-lobe*.

176 **Yuri Gagarin left Earth:** Rebecca Sohn, "Yuri Gagarin: Facts about the First Human in Space," Space.com, January 11, 2024, *https://www.space.com/16159-first-man-in-space.html*.

6. NO MAN'S LAND: LATITUDE

183 **"ticking of the geological clock":** Aldo Leopold, *A Sand County Almanac* (Oxford University Press, 2001), Marshland Elegy, *https://gladcanlit.wordpress.com/wp-content/uploads/2015/02/marshland-elegy.pdf.*

184 **killed more than 2 million people:** Aaron O'Neill, "The Korean War—Statistics & Facts," Statista, August 26, 2024, *https://www.statista.com/topics/10172/the-korean-war/#topicOverview.*

184 **In the wake of this grueling campaign:** The Geographer, Office of the Geographer, "Korea 'Military Demarcation Line' Boundary," *International Boundary Study* no. 22 (May 24, 1963), US Department of State, *https://library.law.fsu.edu/Digital-Collections/LimitsinSeas/pdf/ibs022.pdf.*

184 **160-mile-long demilitarized zone:** "The DMZ and North Korea: What Is the DMZ?," Liberty in North Korea, *https://libertyinnorthkorea.org/blog/the-dmz-and-north-korea.*

185 **rebuilt from the ground up:** Bruce Cumings, historian and author of *Origins of the Korean War*, email, February 10, 2025.

185 **30 million people to more than 78 million:** "South Korea Population," 2024, Worldometer, *https://www.worldometers.info/world-population/south-korea-population/.*; "North Korea Population," 2024, Worldometer, *https://www.worldometers.info/world-population/north-korea-population/.*

185 **survive on corn husks, grass, or tree bark:** Keith B. Richburg, "North Korea on Brink of New Crisis," *The Washington Post*, October 17, 1997, *https://www.washingtonpost.com/archive/politics/1997/10/18/north-korea-on-brink-of-new-crisis/08b790d2-55d1-413f-ab7d-2d62cfb446e1/*; Erin Blakemore, "North Korea's Devastating Famine," *History*, last updated January 27, 2025, *https://www.history.com/news/north-koreas-devastating-famine*; Nicholas Keung, "Exhibit Sheds Light on Brutal Life in North Korea," *The Star*, November 20, 2011, *https://www.thestar.com/news/investigations/exhibit-sheds-light-on-brutal-life-in-north-korea/article_477e220b-ee2d-5273-a0f8-7c5057eeb7a9.html.*

186 **wild boars, mountain goats, Asiatic black bears, and musk deer:** M. Ray Allen, "The Uninhabited DMZ Remains the Most Dangerous Area on Earth," *The Virginian Review*, July 24, 2024, *https://virginianreview.com/227832/*; "Animals Living in the DMZ," South Korean National Institute of Ecology, Google Arts and Culture, *https://artsandculture.google.com/story/twVRv8x4gE_fWw.*

186 **eagles intentionally drop their prey:** Author notes, in-person travel to CCZ with DMZ Ecology Research Institute team, December 8, 2023.

187 **surrounded by machine guns and barbed wire:** George Archibald, co-founder, International Crane Foundation, phone call, July 11, 2024.

187 **like drywood termites:** Rudolf Scheffrahn, professor of entomology, Fort Lauderdale Research and Education Center, University of Florida, email, December 10, 2024.

187 **gray wolf packs:** Kent Laudon, senior environmental scientist specialist—wolf, California Department of Fish and Wildlife, email, December 11, 2024.

187 **biggest modern-day driver of wildlife displacement:** Max Roser, "Why Is Improving Agricultural Productivity Crucial to Ending Global Hunger and Protecting the World's Wildlife?," Our World in Data, March 4, 2024, *https://ourworldindata.org/agricultural-productivity-crucial*.

187 **75 percent of the world's land:** "Media Release: Nature's Dangerous Decline 'Unprecedented'; Species Extinction Rates 'Accelerating,'" Intergovernmental Science-Policy Platform on Biodiversity and Ecosystem Services, May 6, 2019, *https://www.ipbes.net/news/Media-Release-Global-Assessment*.

187 **less room for others:** Catrin Einhorn and Lauren Leatherby, "Animals Are Running Out of Places to Live," *The New York Times*, December 9, 2022, *https://www.nytimes.com/interactive/2022/12/09/climate/biodiversity-habitat-loss-climate.html*.

187 **become increasingly compressed:** Neil A. Gilbert et al., "Human Disturbance Compresses the Spatiotemporal Niche," *Proceedings of the National Academy of Sciences* 119, no. 52 (December 27, 2022): e2206339119, *https://doi.org/10.1073/pnas.2206339119*.

188 **Great American Biotic Interchange:** "Great American Interchange," Wikipedia, *https://en.wikipedia.org/wiki/Great_American_Interchange*.

188 **South American native ungulates:** Dane Pavitt, "The Great American Interchange—When Worlds Collide," *The Average Scientist*, May 9, 2023, *https://theaveragescientist.co.uk/2023/05/09/the-great-american-interchange-when-worlds-collide/*.

188 **meteorologists call "bioscatter":** Matthew Van Den Broeke, associate professor, Department of Earth and Atmospheric Sciences, University of Nebraska—Lincoln, email, December 9, 2024.

190 **Since 2002, he has spent:** Author notes, in-person visit to CCZ with Lee Ki-sup, Waterbird Network Korea, December 9, 2023.

191 **has dwindled in recent decades:** "South Korea Land Use: Land Area," CEIC Data, *https://www.ceicdata.com/en/korea/environmental-land-use-oecd-member-annual/land-use-land-area*.

191 **Paju, the region:** Paju, South Korea, population 2024, World Population Review, *https://worldpopulationreview.com/cities/south-korea/paju*.

191 **more than 800,000 people:** Gimpo, South Korea, population 2024, World Population Review, *https://worldpopulationreview.com/cities/south-korea/gimpo*.

192 **[Map of crane observations]:** Cranes: Lee Ki-sup, based on independent surveys conducted with support from South Korea's National Institute of Ecology. Human population density: Kontur Population (CC BY) *https://data.humdata.org/dataset/kontur-population-dataset (2023)*. CCZ boundaries estimated based on available data resources.

193 **nearly 600 pounds:** Rolf Potts, "Korea's No-Man's-Land," *Salon*, February 3, 1999, *https://www.salon.com/1999/02/03/feature_115/*.

193 **it hangs limp:** Kaushik Patowary, "The Strange Case of Kijong-dong and Daeseong-dong," Amusing Planet, August 13, 2018, *https://www.amusingplanet.com/2018/08/the-strange-case-of-kijong-dong-and.html*.

194 **peer-reviewed 2018 study:** Simone Pika et al., "Taking Turns: Bridging the Gap between Human and Animal Communication," *Proceedings of the Royal Society B: Biological Sciences* 285, no. 1880 (June 6, 2018): 20180598, *https://doi.org/10.1098/rspb.2018.0598*.

194 **be two-way conversations:** Ingrid Newkirk and Gene Stone, *Animalkind: Remarkable Discoveries about Animals and the Remarkable Ways We Can Be Kind to Them* (Simon & Schuster, 2020), 59.

195 **underwent a winter transmutation:** Lucy Cooke, "3 Bizarre (and Delightful) Ancient Theories about Bird Migration," TED-Ed, *https://ed.ted.com/lessons/3-bizarre-and-delightful-ancient-theories-about-bird-migration-lucy-cooke*.

195 ***pfeilstörche*, or arrow storks:** Dr. Lucy Cooke, zoologist, email, August 14, 2024.

196 **tiny arctic terns:** Newkirk and Stone, *Animalkind*, 19.

196 **affixed to Borzya:** Dr. Nyamba Batbayar, director, Wildlife Science and Conservation Center of Mongolia, video call, June 24, 2024.

196 **power line collisions, habitat loss, and poisoning:** Claire Mirande, Nyambayar Batbayar, and James T. Harris, "Species Review: White-Naped Crane (*Grus vipio*)," International Crane Foundation, *https://savingcranes.org/wp-content/uploads/2024/10/crane_conservation_strategy_white-naped_crane.pdf*; Yifei Jia et al., "Shifting of the Migration Route of White-Naped Crane (*Antigone vipio*) Due to Wetland Loss in China," *Remote Sensing* 13, no. 15 (2021), *https://doi.org/10.3390/rs13152984*; Nyambayar Batbayar, "The Journey of Borzya the White-Naped Crane," International Crane Foundation, *https://savingcranes.org/news/resources/the-journey-of-borzya-the-white-naped-crane*.

197 **America's Rust Belt:** Sarah Gibbens, "Birds Once Turned Black from Pollution—What That Teaches Us Today," *National Geographic*, October 10, 2017, *https://www.nationalgeographic.com/science/article/birds-air-pollution-soot-black-carbon-spd*.

197 **weakening their immune system:** Kenneth Qin, "Birds Suffer from Air Pollution, Just like We Do," *Audubon*, July 23, 2015, *https://ca.audubon.org/news/birds-suffer-air-pollution-just-we-do*.

197 **biologging update from Mongolia:** Batbayar, video call, June 24, 2024.

198 **"Twenty-two million people":** Lee Ji-hye, "Greater Seoul Area Accounts for More Than 50% of S. Korea's Population," *Hankyoreh*, July 30, 2021, *https://english.hani.co.kr/arti/english_edition/e_national/1005930.html*.

200 **bisected by the Sami:** Lee Ki-sup, video call, July 22, 2024.

202 **as the "Green Belt":** Christian Schwägerl, "Along Scar from Iron Cur-

tain, a Green Belt Rises in Germany," *Yale Environment 360*, April 4, 2011, *https://e360.yale.edu/features/along_scar_from_iron_curtain_a_green_belt_rises_in_germany*.

202 **Amur River border:** Anna Barma, "Amur Green Belt," University of Montana Center for Natural Resources and Environmental Policy, *https://naturalresourcespolicy.org/docs/hands-across-borders/TBC%20Profiles/tbc-profile-template_amur-green-belt_barma.pdf*.

202 **South American jungles:** Federico Rios, "A Bird-Watchers Paradise, Opened Up by Colombia's Peace Deal," *The New York Times*, June 25, 2024, *https://www.nytimes.com/card/2024/06/25/world/americas/colombia-bird watching*.

202 **more than two hundred thousand people:** Madison Whipple, "A Generation Later, What Does Chernobyl Look Like Today?," The Collector, May 19, 2023, *https://www.thecollector.com/chernobyl-today/*.

203 **resettled there in our wake:** Germán Orizaola, "With Humans Out of the Way, Chernobyl's Wildlife Thrives," *Popular Science*, May 11, 2019, *https://www.popsci.com/chernobyl-refuge-for-wildlife/*.

203 **new level of significance:** Archibald, phone call, July 11, 2024.

204 **increasingly common in South Korea:** The Korea Herald/Asia News Network, "The Rise of Staff-Free Stores in South Korea," Inquirer.net, November 15, 2021, *https://business.inquirer.net/334431/the-rise-of-staff-free-stores-in-south-korea*.

204 **lowest birthrates in the world:** Jessie Yeung, Alex Stambaugh, and Yoonjung Seo, "South Korea's Birth Rate Is So Low, the President Wants to Create a Ministry to Tackle It," CNN, May 9, 2024, *https://www.cnn.com/2024/05/09/asia/south-korea-government-population-birth-rate-intl-hnk/index.html*.

204 **columnist Ross Douthat:** Ross Douthat, "Is South Korea Disappearing?," *The New York Times*, December 2, 2023, *https://www.nytimes.com/2023/12/02/opinion/south-korea-birth-dearth.html*.

206 **new highway across it began in 1991:** Nial Moores, "South Korea's Shorebirds: A Review of Abundance, Distribution, Threats and Conservation Status," *Stilt* 50 (January 1, 2006): 62–72, *https://www.eaaflyway.net/wp-content/uploads/2017/12/NialMoores-SK-shorebirds-2006.pdf*.

206 **a "phantom road":** Ed Yong, "How Animals Perceive the World," *The Atlantic*, June 13, 2022, *https://www.theatlantic.com/magazine/archive/2022/07/light-noise-pollution-animal-sensory-impact/638446/*.

207 **The iron horse wants to run again:** "Woljeong-ri Station," Korea Tourism Organization, *https://english.visitkorea.or.kr/svc/whereToGo/locIntrdn/rgnContentsView.do?vcontsId=106820&menuSn=351*.

207 **railroad island since the '50s:** Kyunghee Park and Youkyung Lee, "Trump-Kim Summit Rekindles Dream of Korean Rail Link with Asia," *Bloomberg*, February 21, 2019, *https://www.bloomberg.com/news*

/articles/2019-02-21/trump-kim-summit-rekindles-dream-of-korean-rail-link-with-asia?embedded-checkout=true.

209 **Recapture Sabseulbong at all costs:** Author notes, exhibition video during in-person visit to Crane Ecology Observation, Cheorwon, South Korea, December 2, 2023.

211 **average of twenty to thirty years:** Archibald, phone call, July 11, 2024.

211 **stopped coming farther south in recent years:** Yuko Haraguchi, curator, Crane Park Izumi, email, June 25, 2024.

211 **one thousand red-crowned**: Kim Hae-yeon, "Birdwatching near Demilitarized Zone," *The Korea Herald*, January 7, 2022, *https://www.koreaherald.com/article/2754316.*

211 **ten thousand white-naped cranes:** "Nature Rises from War Ravages," *The Business Times*, October 7, 2017, *https://www.businesstimes.com.sg/incoming/nature-rises-war-ravages.*

EPILOGUE

220 **based on the tracks of elk and other nonhuman animals:** Dennis Garcia and Jewlya Samaniego, Chumash community members, phone calls, July 16, 2024.

220 **christened El Camino Real:** "History of El Camino," Grand Boulevard Initiative, *https://grandboulevard.net/about/history-of-el-camino.*

220 **call an "extinction vortex":** Lauren Gill, California deputy director, National Wildlife Federation, video call, August 12, 2024.

220 **about a deceased animal:** City News Service, "Dead Mountain Lion Found on 101 Freeway near Wildlife Crossing Site," *Los Angeles Daily News*, June 15, 2024, *https://www.dailynews.com/2024/06/15/dead-mountain-lion-found-on-101-freeway-near-wildlife-crossing-site/.*

221 **two "biodiversity hotspots":** "Crossing FAQs," Wallis Annenberg Wildlife Crossing, *https://101wildlifecrossing.org/crossing-faq/.*

221 **cougar-free within fifteen years:** Gill, video call, August 12, 2024.

221 **2,700 square miles of Los Padres National Forest beckon:** Gill, video call, August 12, 2024.

221 **football field's worth:** "Crossing FAQs," Wallis Annenberg Wildlife Crossing, *https://101wildlifecrossing.org/crossing-faq/.*

221 **other native plants:** Samaniego, in-person interview, August 13, 2024.

221 **largest wildlife crossing anywhere in the world:** "World's Largest Wildlife Crossing on Track to Open by Early 2026," Governor Gavin Newsom, May 7, 2024, *https://www.gov.ca.gov/2024/05/07/worlds-largest-wildlife-crossing-on-track-to-open-by-early-2026/.*

221 **mammoth 210-foot:** "Crossing FAQs," Wallis Annenberg Wildlife Crossing, *https://101wildlifecrossing.org/crossing-faq/.*

221 **"vegetated bridge" that will:** "US-101—Wallis Annenberg Wildlife Crossing at Liberty Canyon," California Department of Transportation (Caltrans), *https://dot.ca.gov/caltrans-near-me/district-7/district-7-projects/d7-101-annenberg-wildlife-crossing*.

221 **coyotes, bobcats, birds, frogs, butterflies, and lizards:** "Crossing FAQs," Wallis Annenberg Wildlife Crossing, *https://101wildlifecrossing.org/crossing-faq/*.

222 **[Wildlife crossing sketch]:** While this book's illustrator, Oliver Uberti, drew most of his illustrations after I returned from the field, he was able to join me for a few adventures, including a midnight hike to view the construction of the Wallis Annenberg Wildlife Crossing. Fully aware that cougars roam these hills, Oliver, ever the perfectionist, insisted the risk was worth it to ensure he got the lighting and color of the scene just right.

224 **will need soundproof walls:** Tara Lynn Wagner, "Sound Barriers Help Wallis Annenberg Wildlife Crossing Create a Calm Environment above the Roar Traffic," *Spectrum News 1*, October 15, 2024, *https://spectrumnews1.com/ca/southern-california/environment/2024/10/16/sound-barriers-help-wallis-annenberg-wildlife-crossing-create-a-calm-environment-above-the-roar-traffic*.

224 **2,700 tons of steel and 39,000 tons of concrete:** Lawrence Okoye, PE, senior bridge engineer, Caltrans, email, August 28, 2024.

226 **without touching concrete once:** Gill, email, September 10, 2024.

226 **toads to successfully cross:** Beth Pratt (@bethpratt), "WE GOT TO SEE A YOSEMITE TOAD USE A WILDLIFE CROSSING!," X, August 21, 2024, *https://x.com/bethpratt/status/1826484644390601033*.

226 **Chileno Valley Newt Brigade:** Annie Roth, "A 'Big Night' for Newts, and for a California Newt Brigade," *The New York Times*, January 24, 2023, *https://www.nytimes.com/2023/01/24/science/newts-roads-california.html*.

226 **underpass below the 118 freeway:** "Caltrans and NPS Retrofit Project Helps Wildlife Cross Highway 118," National Park Service, April 14, 2021, *https://www.nps.gov/samo/learn/news/caltrans-and-nps-retrofit-project-helps-wildlife-cross-highway-118.htm*.

227 **by up to 97 percent:** "Crossing FAQs," Wallis Annenberg Wildlife Crossing, *https://101wildlifecrossing.org/crossing-faq/*.

227 **In Utah, more than seven hundred animals:** Parker Malatesta, "Over 700 Animals Used the I-80 Wildlife Bridge in 2021," *TownLift*, January 3, 2022, *https://townlift.com/2022/01/over-700-animals-used-the-i-80-wildlife-bridge-in-2021/*.

227 **just west of Denver:** Brian Sherrod, "New Wildlife Underpass at I-70 West of Denver Designed to Keep More Animals, Drivers Safe," CBS Colorado, June 20, 2024, *https://www.cbsnews.com/colorado/news/new-wildlife-underpass-i-70-west-denver-designed-animals-drivers-safe/*.

227 **Across the Mountain West:** Starre Vartan, "How Wildlife Bridges over

Highways Make Animals—and People—Safer," *National Geographic*, April 16, 2019, *https://www.nationalgeographic.com/animals/article/wildlife-overpasses-underpasses-make-animals-people-safer*.

227 **system of "fish ladders":** "Fish Tale," *National Geographic*, *https://education.nationalgeographic.org/resource/fish-tale/*.

227 **up to a thousand miles:** Michael Collins, "Pacific Salmon Travel Thousands of Miles, Return Home, and Feed the Forest," Teravana, n.d., *https://teravana.org/pacific-salmon-travel-thousands-of-miles/*.

227 **In Belgium, a bridge:** Jules Johnston, "Animals Spotted Crossing Brussels Ring Road 'Eco Bridge,'" *The Brussels Times*, July 30, 2019, *https://www.brusselstimes.com/61894/brussels-animals-recorded-crossing-ring-road-eco-bridge*.

227 **Wild boars in Germany:** Nicole Glass, "Green Bridges Help Animals Safely Cross the German Autobahn," GermanyinUSA, May 13, 2021, *https://germanyinusa.com/2021/05/13/green-bridges-help-animals-safely-cross-the-german-autobahn/*.

227 **lynxes in the Czech Republic:** Balkan Green Energy News, "Green Bridges across Roads and Highways Saving Bears, Lynxes, Wolves," Balkan Green Energy News, May 27, 2020, *https://balkangreenenergynews.com/green-bridges-across-roads-and-highways-saving-bears-lynxes-wolves/*.

227 **red squirrels in the UK:** Richard Bunting, "New 'Suspension Bridge' Keeps Red Squirrels Safe in Highlands," Trees for Life, n.d., *https://treesforlife.org.uk/new-lsquosuspension-bridgersquo-keeps-red-squirrels-safe-in-highlands/*.

227 **tigers in India:** Gana Kedlaya, "Monitoring Crossing Structures in Tiger Reserves Offers Insights into Species Movement, Behaviour," *Mongabay*, July 25, 2022, *https://india.mongabay.com/2022/07/monitoring-crossing-structures-in-tiger-reserves-offers-insights-into-species-movement-behaviour/*.

227 **koalas and possums in Australia:** "Koala Crossing," Fauna Crossings, *https://faunacrossings.com.au/koala-crossings/*.

227 **macaques in Singapore:** Navene Elangovan, "Saving Singapore's Endangered Species, One 'Animal Bridge' at a Time," *Today*, February 12, 2022, *https://www.todayonline.com/big-read/big-read-saving-singapores-endangered-species-one-animal-bridge-time-1814596*.

227 **on Christmas Island:** Mark Sawa, public affairs, Parks Australia, email, August 7, 2024.

228 **[Crab bridge image]:** Wondrous World Images—Yvonne McKenzie, republished with permission, August 25, 2024.

229 **43 miles from the beach:** Captain Garrick Hyder, owner-operator, Santa Barbara Boat Charters, phone call, September 8, 2024.

230 **known as *tomols*:** "Chumash Tomol Crossing," National Park Service, *https://www.nps.gov/chis/learn/historyculture/tomolcrossing.htm*.

230 **San Miguel has been "uninhabited":** "Uninhabited San Miguel Island

Reopening to Visitors, Campers," KCAL News, May 4, 2016, *https://www.cbsnews.com/losangeles/news/uninhabited-san-miguel-island-reopening-to-visitors-campers/*.

230 **since the 1940s:** John Woodard and Larry Goldman, volunteers, Channel Islands Visitors Center, phone calls, September 20, 2024, and April 15, 2025.

230 **presumed to be largely empty:** Dino Grandoni, "Scientists Detected 5,000 Sea Creatures Nobody Knew Existed. It's a Warning," *The Washington Post*, May 25, 2023, *https://www.washingtonpost.com/climate-environment/2023/05/25/deep-sea-mining-environmental-impacts/*.

230 **giant tube worms:** Craig R. McClain et al., "Sizing Ocean Giants: Patterns of Intraspecific Size Variation in Marine Megafauna," *PeerJ* 3, no. 715 (January 13, 2015), *https://pmc.ncbi.nlm.nih.gov/articles/PMC4304853/*.

230 **previously unknown species:** Muriel Rabone et al., "How Many Metazoan Species Live in the World's Largest Mineral Exploration Region?," *Current Biology* 33, no. 12 (June 19, 2023): 2383–2396.e5, *https://doi.org/10.1016/j.cub.2023.04.052*.

230 **"unmapped, unobserved, and unexplored":** "How Many Species Live in the Ocean?," National Ocean Service, National Oceanic and Atmospheric Administration, *https://oceanservice.noaa.gov/facts/ocean-species.html*.

230 **a million species:** "How Much of the Ocean Has Been Explored?," National Oceanic and Atmospheric Administration, *https://oceanexplorer.noaa.gov/facts/explored.html*.

231 **not the octopuses:** Ephrat Livni, "Octlantis Is a Just-Discovered Underwater City Engineered by Octopuses," Quartz, September 17, 2017, *https://qz.com/1077632/octlantis-is-a-just-discovered-underwater-city-engineered-by-octopuses*.

231 **beluga whale calls:** "Using Artificial Intelligence to Identify Endangered Beluga Whales," NOAA Fisheries, March 24, 2020, *https://www.fisheries.noaa.gov/feature-story/using-artificial-intelligence-identify-endangered-beluga-whales*.

231 **more than 600,000 square miles:** "Frequently Asked Questions," Chumash Heritage National Marine Sanctuary, *https://chumashsanctuary.org/faq/*.

232 **4,543 square miles:** "Rep. Carbajal Celebrates Finalization of Chumash Heritage National Marine Sanctuary," U.S. Congressman Salud Carbajal, December 2, 2024, *https://carbajal.house.gov/news/documentsingle.aspx?DocumentID=3084*.

232 **"Serengeti of the Sea":** 805 Aerial, "Chumash Heritage National Marine Sanctuary: Insights," posted on January 28, 2017, Vimeo, 5 min., 10 sec., timestamp: 3 min., 50 sec., *https://vimeo.com/201404320*.

232 **5,000 nautical miles round-trip:** Dr. Mark O. Lammers, research ecolo-

gist, Hawaiian Islands Humpback Whale National Marine Sanctuary, email, August 19, 2024.

232 **restricted or prohibited completely:** Lammers, email, August 19, 2024.

232 **trophies to the shipping companies:** Cara Buckley, "And the Winner Is . . . the Slowest!," *The New York Times*, July 2, 2024, *https://www.nytimes.com/2024/07/02/climate/whales-speed-ships-slow.html*.

232 **one hundred thousand individuals have already been conclusively identified:** Emily Anthes, "How the World's Oldest Humpback Whale Has Survived Is a Mystery," *The New York Times*, August 14, 2024, *https://www.nytimes.com/2024/08/14/science/oldest-humpack-whale-old-timer.html*.

233 **renowned for its shipwrecks:** "Shipwrecks," Channel Islands National Park, National Park Service, *https://www.nps.gov/places/000/shipwrecks.htm*.

233 **more than 16 feet long:** "Robinson, James," Islapedia, *https://www.islapedia.com/index.php?title=ROBINSON,_James*.

234 **southern terminus for northern fur seals:** Woodard, phone call, September 20, 2024.

234 **all the way from the Pribilof Islands:** Dr. Tony Orr, wildlife biologist, Marine Mammal Laboratory, National Oceanic and Atmospheric Administration, email, September 20, 2024.

235 **45 percent of the global breeding population:** Tony Orr, "Studying Fur Seal and Sea Lion Populations in Sunny Southern California—Blog Post 1," NOAA Fisheries, July 23, 2024, *https://www.fisheries.noaa.gov/science-blog/studying-fur-seal-and-sea-lion-populations-sunny-southern-california-blog-post-1*.

235 **visit San Miguel each year:** "Point Bennett," Channel Islands National Park, National Park Service, *https://www.nps.gov/places/000/point-bennett.htm*.

236 **largest Chumash village:** "Chumash on San Miguel Island," Channel Islands National Park, National Park Service, *https://www.nps.gov/places/000/chumash-on-san-miguel-island.htm*.

236 **subspecies of island fox:** "Island Fox," Channel Islands National Park, National Park Service, *https://www.nps.gov/chis/learn/nature/island-fox.htm*.

237 **radiant, sustaining energy:** Wilding, email, July 8, 2024.

237 **"There were times, on the threshold of spring":** Henry Beston, *The Outermost House* (Penguin Books, 1988), 94–95.

About the Author

RYAN HULING is an explorer and sustainable food systems specialist based in Sierra Madre, California. He currently serves as a senior writer at the Good Food Institute Asia Pacific. His columns have appeared in *Nature*, *Wired*, *USA Today*, *The Nikkei*, and the *South China Morning Post*.

About the Illustrator

OLIVER UBERTI is a former senior design editor for *National Geographic* and the coauthor and designer of four acclaimed books of maps and graphics, including *Atlas of Finance*, which *The Wall Street Journal* called "a master class in the use of design as exposition." His books, maps, and data visualizations have earned many honors, including the Corlis Benefideo Award for Imaginative Cartography from the North American Cartographic Information Society, the Wenjin Book Award from the National Library of China, and the Geographical Engagement Award from the Royal Geographical Society. At his studio in Los Angeles, Oliver continues to help authors and scientists translate their research into memorable visuals that illuminate the wonder of the world.